T0181581

Studies in Fuzziness and Soft Computing   300

Editor-in-Chief

Prof. Janusz Kacprzyk
Systems Research Institute
Polish Academy of Sciences
ul. Newelska 6
01-447 Warsaw
Poland
E-mail: kacprzyk@ibspan.waw.pl

For further volumes:
http://www.springer.com/series/2941

Michał Baczyński, Gleb Beliakov,
Humberto Bustince Sola, and Ana Pradera (Eds.)

Advances in Fuzzy
Implication Functions

 Springer

Editors

Michał Baczyński
Institute of Mathematics
University of Silesia
Katowice
Poland

Gleb Beliakov
School of Information Technology
Deakin University
Burwood
Australia

Humberto Bustince Sola
Departamento de Automática y
Computación
Universidad Pública de Navarra
Pamplona
Spain

Ana Pradera
Departamento de Ciencias de la
Computación
Universidad Rey Juan Carlos
Madrid
Spain

ISSN 1434-9922   e-ISSN 1860-0808
ISBN 978-3-642-43779-3   ISBN 978-3-642-35677-3 (eBook)
DOI 10.1007/978-3-642-35677-3
Springer Heidelberg New York Dordrecht London

Preface

This volume collects research papers on fuzzy implication functions, which are fundamental for fuzzy logic systems, fuzzy control, decision theory, expert systems etc. They play also a key role in fuzzy mathematical morphology and in solving fuzzy relation equations. Based on the evaluations of our referees, we have accepted 8 manuscripts. These papers aim to present today's state-of-the-art in this area.

This volume starts with the work by Massanet and Torrens ("An Overview of Construction Methods of Fuzzy Implications"), which recalls several distinct types of fuzzy implications and their properties, and then focuses on the construction methods. The authors discuss implications generated from one or more initial implications, and implications generated from aggregation functions or fuzzy negations.

It is followed by the work of Shi, Van Gasse, and Kerre ("Fuzzy Implications: Classification and a New Class"), which focuses on the common properties of implication functions, their dependencies, and on methods of generating new classes of implications. One new class is presented, based on fuzzy negation.

The article by Qin and Baczyński ("A Survey of the Distributivity of Implications over Continuous T-norms and the Simultaneous Satisfaction of the Contrapositive Symmetry") summarizes the sufficient and necessary conditions of solutions for the distributivity equation of implication $I(x, T_1(y, z)) = T_2(I(x, y), I(x, z))$ when T_1 is a continuous triangular norm, T_2 is a continuous Archimedean triangular norm and I is an unknown function. The authors characterize also all solutions of the system of functional equations consisting of the previous equation and $I(x, y) = I(N(y), N(x))$, where N is a strong negation. Presented methods can be applied to other distributivity functional equations of implications.

The following two papers discuss implications defined on lattices, in particular implications in interval-valued fuzzy set theories. The paper by Deschrijver ("Implication Functions in Interval-valued Fuzzy Set Theory") gives an overview of possible extensions of fuzzy implications to interval-valued case, and gives a characterization of such implications that satisfy Smets-Magrez axioms. It also discusses distributivity of implication functions over triangular norms and triangular conorms (in interval-valued fuzzy set theory) and mentions a few new constructions.

Bedregal, Beliakov, Bustince, Fernandez, Pradera and Reiser ("(S,N)-Implications on Bounded Lattices") study natural extension of (S,N)-implications to arbitrary bounded lattices, applicable to interval-valued, Atanassov intuitionistic fuzzy logics, and fuzzy multiset-based logics. This paper focuses in particular on various properties of implications on bounded lattices. While most properties follow analogously to the $[0, 1]$-valued case, some properties must be weakened, so they are not completely equivalent to the usual fuzzy implications.

The paper by Hliněná, Kalina, and Král' ("Implication Functions Generated Using Functions of one Variable") presents a survey of construction methods based on a single variable generating function. This type of construction mimics generation of the Archimedean t-norms and t-conorms via additive generators. In fact, additive generators of some t-norms and t-conorms have been used to generate $f-$ and $g-$ fuzzy implications. Several other new construction methods are also reviewed and multiple examples are presented.

Drewniak and Sobera ("Compositions of Fuzzy Implications") analyze compositions based on a binary operation $*$ and discuss the dependencies between the algebraic properties of this operation and the induced sup-* composition. Under some assumptions, the $\sup -*$ composition of fuzzy implications also gives a fuzzy implication. Specific attention is paid to ordered algebraic structures of fuzzy implications.

In the last article, Baczyński and Jayaram ("Fuzzy Implications: Some Recently Solved Problems") discuss some open problems related to fuzzy implications, which have either been completely solved or those for which partial answers are known. The recently solved problems are so chosen to reflect the importance of the problem or the significance of the solution. Some other problems that still remain unsolved are stated for quick reference.

As editors we wish to thank all the contributors for their excellent work. All the manuscripts were anonymously peer reviewed by at least two reviewers, and we also wish to express our thanks to them. We hope that the volume will be interesting and useful to the entire intelligent systems research community. We also wish to thank Prof. Janusz Kacprzyk and Dr. Thomas Ditzinger from Springer for their support and encouragement.

November 2012

Michał Baczyński
Gleb Beliakov
Humberto Bustince Sola
Ana Pradera

Contents

An Overview of Construction Methods
of Fuzzy Implications

Sebastià Massanet and Joan Torrens

Abstract. Fuzzy implications are useful in a wide range of applications. For these practical purposes, different classes of implications are used. Depending on the concrete application the implication is going to perform, several additional properties have to be fulfilled. In this paper, we recall briefly the most used classes of implications, (S, N) and R-implications, QL and D-operations, their generalizations to other aggregation functions and Yager's implications, and we show the new construction methods presented in recent years. These construction methods vary from implications defined from general aggregation functions or fuzzy negations, to implications generated from one or two initial implications. For every single class, we determine which additional properties are satisfied.

Keywords: Fuzzy implication, construction method, fuzzy negation, exchange principle, contrapositive symmetry.

1 Introduction

Fuzzy implications have become one of the main operations in fuzzy logic, playing a similar role to the classical implication in crisp logic. It is well-known that the fuzzy concept has to generalize the corresponding crisp one, and consequently fuzzy implications restricted to $\{0, 1\}^2$ must coincide with the classical implication. Nowadays, it is well settled that fuzzy implications are performed by means of a binary operation $I : [0, 1]^2 \rightarrow [0, 1]$ satisfying, in addition of that boundary conditions, left antitonicity and right isotonicity. The decreasingness in the first variable incorporates the idea that a lower truth value of the first variable is more efficient to state

Sebastià Massanet · Joan Torrens
Department of Mathematics and Computer Science
University of the Balearic Islands
Crta. Valldemossa Km. 7.5, E-07122 Palma de Mallorca, Spain
e-mail: {s.massanet,jts224}@uib.es

M. Baczyński et al. (Eds.): *Adv. in Fuzzy Implication Functions*, STUDFUZZ 300, pp. 1–30.
DOI: 10.1007/978-3-642-35677-3_1   © Springer-Verlag Berlin Heidelberg 2013

more about the truth value of its consequent. On the other hand, the increasingness in the second variable is connected to the idea that the overall truth value depends on the consequent directly. Thus, some restrictions are imposed to binary functions in order to be fuzzy implications but they are flexible enough to allow several classes of implications with different additional properties.

However, although theoretically different classes of fuzzy implications are possible, the question that arises immediately is related to the necessity of having so many different classes. The main reason is because fuzzy implications are used to perform any fuzzy "if-then" rule in fuzzy systems and inference processes, through Modus Ponens and Modus Tollens (see [22]). So, depending on the context, and on the proper rule and its behavior, different implications with different properties can be adequate. This is also true in other fields where fuzzy implications play an important role, such as fuzzy mathematical morphology [23] or fuzzy DI-subsethood measures and image processing [12, 13], among many others.

The logical consequence of this fact is the proposal of several classes of implications. Among the most used ones, we can highlight the class of (S,N)-implications [4], that includes the well-known S-implications [40], residual or R-implications [6, 18], QL-operations and D-operations [25]. From these classes, some generalizations based on considering now uninorms [7, 15, 26], copulas or quasi-copulas [16] and co-copulas [42] were proposed generating new classes of implications with interesting properties. Note that all these classes of implications are generated from aggregation functions [19]. However, there exists a different approach in order to obtain fuzzy implications based on the direct use of additive generating functions. In this way, Yager's f- and g-generated fuzzy implications [41] can be seen as implications generated from continuous additive generators of continuous Archimedean t-norms or t-conorms, respectively. Analogously, Balasubramaniam's h-generated implications [8, 9] can be seen as implications generated from multiplicative generators of t-conorms. Most of these classes are notably collected and studied in [27] and [5].

One step further in the generation of fuzzy implications includes those methods that generate a new fuzzy implication from some initial implication(s). The preservation of the additional properties satisfied by the initial implication(s) to the generated one is a desirable fact of these generation methods. The most important of these methods is the conjugation of a fuzzy implication [3], which preserves almost all the properties. Other methods are the N-reciprocation (Section 1.6 in [5]), the upper, lower and medium contrapositivisation [8, 10, 17], the min or max-generation (Section 6.1 in [5]) and the convex combination of fuzzy implications (Section 6.2 in [5]).

In the last years, more interesting classes of implications have been proposed through the different techniques already mentioned. First of all, between the classes generated from some aggregation functions, (TS,N)-implications generated from TS-functions [11]; R-implications from representable aggregation functions or RAF [14]; S-implications from dual representable aggregation functions or DRAF [1] and R-operations from general aggregation functions [35] have joined the existing classes. Furthermore, using the additive generators of representable uninorms, some generalizations of Yager's implications, called h, (h,e) and generalized h-implications have been introduced in [28] (see Chapters [38, 20] of this book for further details on this class and some more new classes of implications generated from additive generators). In addition, some classes of implications generated only from a fuzzy negation have been proposed in [39]. If this was not enough, the threshold generation method and the analogous vertical one, presented in [31, 34] and [29], respectively, give us new possibilities of generating fuzzy implications from two initial ones. Thus, it is remarkable the exponential number of new classes of implications presented in the last years and consequently, it is adequate and interesting to collect them, study their additional properties and how they are related each other, as we want to do in this chapter.

The chapter is organized as follows. In the next section we recall the basic definitions and properties of implications needed in the subsequent sections. In Section 3, we briefly recall the definitions and basic properties satisfied by the most usual classes of implications. Then, in Section 4, we focus into detail with the recently introduced classes of implications, studying their properties and characterizing some of them. Finally, the chapter ends with some conclusions and future work.

2 Preliminaries

We will suppose the reader to be familiar with the theory of aggregation functions (all necessary results and notations can be found in [19] and [24] for the particular case of t-norms and t-conorms). To make this work self-contained, we recall here some of the concepts and results used in the rest of the paper. First of all, we introduce the notions of automorphism and conjugate.

Definition 1. A function $\varphi : [0,1] \to [0,1]$ is an automorphism if it is continuous and strictly increasing and satisfies the boundary conditions $\varphi(0) = 0$ and $\varphi(1) = 1$, i.e., if it is an increasing bijection in $[0,1]$.

Definition 2. Let $\varphi : [0,1] \to [0,1]$ be an automorphism. We say that two functions $f,g : [0,1]^n \to [0,1]$ are φ-conjugate if $g = f_\varphi$, where

$$f_\varphi(x_1,\ldots,x_n) = \varphi^{-1}(f(\varphi(x_1),\ldots,\varphi(x_n))), \quad x_1,\ldots,x_n \in [0,1].$$

Note that given an automorphism $\varphi : [0,1] \to [0,1]$, the φ-conjugate of a t-norm T, that is T_φ, and the φ-conjugate of an implication I (see Definition 4), that is I_φ, are again a t-norm and an implication, respectively.

2.1 Fuzzy Negations

Definition 3. (see Definition 1.1 in [18]) A decreasing function $N : [0,1] \to [0,1]$ is called a *fuzzy negation*, if $N(0) = 1$, $N(1) = 0$. A fuzzy negation N is called

  (i) *strict*, if it is strictly decreasing and continuous.
 (ii) *strong*, if it is an involution, i.e., $N(N(x)) = x$ for all $x \in [0,1]$.

Example 1. Important negations that will be used throughout this paper are the *standard negation*, $N_C(x) = 1 - x$, the *Sugeno class* of negations $N^\lambda(x) = \frac{1-x}{1+\lambda x}$ with $\lambda \in (-1, +\infty)$, the *least* or Gödel, and the *greatest* or dual Gödel fuzzy negations given respectively by

$$N_{D_1}(x) = \begin{cases} 1 \text{ if } x = 0, \\ 0 \text{ if } x > 0, \end{cases} \quad N_{D_2}(x) = \begin{cases} 1 \text{ if } x < 1, \\ 0 \text{ if } x = 1. \end{cases}$$

2.2 Fuzzy Implications

Definition 4. (Definition 1.15 in [18]) A binary operator $I : [0,1]^2 \to [0,1]$ is said to be an *implication function*, or an *implication*, if it satisfies:

(I1) $I(x,z) \geq I(y,z)$ when $x \leq y$, for all $z \in [0,1]$.
(I2) $I(x,y) \leq I(x,z)$ when $y \leq z$, for all $x \in [0,1]$.
(I3) $I(0,0) = I(1,1) = 1$ and $I(1,0) = 0$.

Note that, from the definition, it follows that $I(0,x) = 1$ and $I(x,1) = 1$ for all $x \in [0,1]$ whereas the symmetrical values $I(x,0)$ and $I(1,x)$ are not derived from the definition. The expressions of the basic fuzzy implications are collected in Table 1.3 in [5]. Some properties of fuzzy implications that will be used throughout this paper are the following:

- The *left neutrality principle*,

$$I(1,y) = y, \quad y \in [0,1]. \tag{NP}$$

- The *exchange principle*,

$$I(x, I(y,z)) = I(y, I(x,z)), \quad x,y,z \in [0,1]. \tag{EP}$$

- The *law of importation* with a t-norm T,

$$I(T(x,y),z) = I(x, I(y,z)), \quad x,y,z \in [0,1]. \tag{LI}$$

- The *ordering property*,

$$x \leq y \Longleftrightarrow I(x,y) = 1, \quad x,y \in [0,1]. \tag{OP}$$

- The *identity principle*,

$$I(x,x) = 1, \quad x \in [0,1]. \tag{IP}$$

- The *contrapositive symmetry* with respect to a fuzzy negation N,

$$I(x,y) = I(N(y), N(x)), \quad x,y \in [0,1]. \tag{CP(N)}$$

Definition 5. (see Definition 1.14.15 in [5]) Let I be a fuzzy implication. The function N_I defined by $N_I(x) = I(x,0)$ for all $x \in [0,1]$, is called the *natural negation* of I.

3 Main Types of Implications

There exist basically two strategies in order to define fuzzy implications. The most general strategy is based on some combinations of aggregation functions. In this way, t-norms and t-conorms were the first classes of aggregation functions used to generate fuzzy implications. Thus, the following are the four usual models of fuzzy implications:

1) (S,N)-implications defined as

$$I(x,y) = S(N(x),y), \quad x,y \in [0,1]$$

where S is a t-conorm and N a fuzzy negation. They are the immediate generalization of the classical boolean material implication $p \to q \equiv \neg p \vee q$. If N is strong, we recover strong or S-implications.

2) Residual or R-implications defined by

$$I(x,y) = \sup\{z \in [0,1] \mid T(x,z) \leq y\}, \quad x,y \in [0,1]$$

where T is a t-norm. When they are obtained from left-continuous t-norms, they come from residuated lattices based on the residuation property

$$T(x,y) \leq z \Leftrightarrow I(x,z) \geq y, \text{ for all } x,y,z \in [0,1].$$

3) QL-operations defined by

$$I(x,y) = S(N(x),T(x,y)), \quad x,y \in [0,1]$$

where S is a t-conorm, T is a t-norm and N is a fuzzy negation. Their origin is the quantum mechanic logic.

4) D-operations defined by

$$I(x,y) = S(T(N(x),N(y)),y), \quad x,y \in [0,1]$$

where S is a t-conorm, T is a t-norm and N is a fuzzy negation. They are the contraposition with respect to N of QL-operations when N is strong. Their name comes from the Dishkant arrow $p \to q \equiv q \vee (\neg p \wedge \neg q)$ of orthomodular lattices.

Note that R and (S,N)-implications are always implications in the sense of Definition 4, whereas QL and D-operations are not implications in general. A characterization of those cases when QL or D-operations are implications is still open (Problem 2.7.5 in [5]). However, in both cases a common necessary condition is $S(N(x),x) = 1$ for all $x \in [0,1]$.

These initial classes have been generalized after considering more general classes of aggregation functions. Thus, copulas, quasi-copulas and even conjunctions in general ([16]), but mainly uninorms ([2, 7, 15, 26]) can be successfully used to generate new classes of fuzzy implications. So, the so-called (U,N) and RU-implications and the QLU and DU-operations were introduced and deeply studied, proving that they satisfy interesting properties, which are the counterparts of the common properties presented in Section 2 for implications derived from uninorms. Some of them are listed below for some $e \in (0,1)$ (usually the neutral element of the involved uninorm):

- The *left neutrality principle for implications derived from uninorms*,

$$I(e,y) = y, \quad y \in [0,1]. \tag{NP_e}$$

- The *law of importation* with a conjunctive uninorm U,

$$I(U(x,y),z) = I(x,I(y,z)), \quad x,y,z \in [0,1]. \tag{LI_U}$$

- The *ordering property for implications derived from uninorms*,

$$x \leq y \Longleftrightarrow I(x,y) \geq e, \quad x,y \in [0,1]. \tag{OP_e}$$

- The *identity principle for implications derived from uninorms*,

$$I(x,x) = e, \quad x \in (0,1). \tag{IP_e}$$

- The *natural negation of I with respect to e* given by

$$N_I^e(x) = I(x,e), \quad x \in [0,1].$$

For further details on these classes of implications, we recommend [5, 26, 27, 36] and [37].

On the other hand, among the implications generated from additive generators, we can highlight Yager's f and g-generated implications which have been recently characterized in [32] and their intersection with QL and D-implications have been fully characterized in [30]. However, there are still some other open problems related to this class of implications such as the characterization of their convex closure (Problem 6.5.1 in [5]) and the study of the T-conditionality property on this class of implications. A deep study on this strategy to generate fuzzy implications can be found in another chapter [38] of this book.

3.1 Methods to Obtain Implications from Old Ones

An interesting fact in the generation of fuzzy implications is that any implication can be itself modified with some method to obtain a new one. The usual methods to obtain implications from some old ones are the following:

- The N-reciprocation of a fuzzy implication I

$$I_N(x,y) = I(N(y),N(x)), \quad x,y \in [0,1],$$

where N is a fuzzy negation. The function I_N is called the N-reciprocal of I.

- The medium contrapositivisation of a fuzzy implication I

$$I_N^m(x,y) = \min\{I(x,y) \vee N(x), I_N(x,y) \vee y\}, \quad x,y \in [0,1],$$

where N is a fuzzy negation.
- The φ-conjugation of a fuzzy implication I

$$I_\varphi(x,y) = \varphi^{-1}(I(\varphi(x),\varphi(y))), \quad x,y \in [0,1],$$

where φ is an automorphism on $[0,1]$.
- The min and max operations generate new fuzzy implications from two given ones I, J:

$$(I \vee J)(x,y) = \max\{I(x,y), J(x,y)\}, \quad x,y \in [0,1],$$

$$(I \wedge J)(x,y) = \min\{I(x,y), J(x,y)\}, \quad x,y \in [0,1].$$

- The convex combinations of fuzzy implications also provides new fuzzy implications from two given ones I, J and $\lambda \in [0,1]$:

$$I_{I,J}^\lambda(x,y) = \lambda \cdot I(x,y) + (1-\lambda) \cdot J(x,y), \quad x,y \in [0,1].$$

There are other methods, such as the lower and upper contrapositivisation but they are combinations of these listed methods. For example, the lower (upper) contrapositivisation is based on applying the min (max) method to a fuzzy implication I and its N-reciprocal.

The relevance of these methods is based on their capability of preserving the additional properties satisfied by the initial implication(s). The preservation of some properties is studied in [5], but since some of them are missing, let us study the remaining ones. Let us start with the N-reciprocation method. This method always preserves (IP), (OP) and the continuity when N is continuous. Then we have that the N-reciprocal implication satisfies (NP) if, and only if, $N_I \circ N = \mathrm{id}_{[0,1]}$, and if N is strong, it is an involution. However, (EP) and (LI) are not generally preserved. Let us study their natural negation.

Proposition 1. *Let I be a fuzzy implication and N a fuzzy negation. The natural negation of I_N is $N_{I_N}(x) = I(1,N(x))$. Moreover,*

(i) If I satisfies (NP), $N_{I_N} = N$.
(ii) If $I(1,\cdot)$ is continuous (and strict), then N_{I_N} is continuous (strict).
(iii) If $I(1,N(I(1,N(x)))) = x$, then N_{I_N} is strong.

Proof. The expression of the natural negation comes from

$$N_{I_N}(x) = I_N(x,0) = I(N(0),N(x)) = I(1,N(x)).$$

Now, (i) and (ii) are immediate. Finally, if $I(1,N(I(1,N(x)))) = x$, then we have

$$N_{I_N}(N_{I_N}(x)) = I(1,N(N_{I_N}(x))) = I(1,N(I(1,N(x)))) = x.$$

So, N_{I_N} is a strong negation. $\qquad\square$

Next, the medium contrapositivisation preserves (NP), (OP) and (IP). In addition, if N is strong, then I_N^m satisfies (CP(N)) and $(I_N^m)_N^m = I_N^m$. Finally, it is easy to show that it does not preserve either (EP) or (LI).

Example 2. Let us consider the Yager implication I_{YG} and its N_C-reciprocal. We know that I_{YG} satisfies (EP), but its medium contrapositivisation with respect to N_C (see Figure 1) given by

$$I_{N_C}^m(x,y) = \begin{cases} 1 & \text{if } x = y = 1 \text{ or } x = y = 0, \\ \min\{\max\{y^x, 1-x\}, \max\{(1-x)^{(1-y)}, y\}\} & \text{otherwise,} \end{cases}$$

does not satisfy (EP), since

$$I_{N_C}^m(0.3, I_{N_C}^m(0.7, 0.3)) = 0.776597 \neq 0.77 = I_{N_C}^m(0.7, I_{N_C}^m(0.3, 0.3)).$$

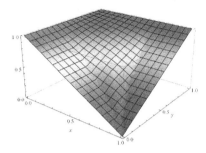

Fig. 1 Counterexample of the preservation of (EP) by the medium contrapositivisation method

On the other hand, the φ-conjugation of a fuzzy implication preserves (NP), (EP), (OP) and (IP). In addition, the natural negation of a φ-conjugate of an implication I is given by $N_{I_\varphi} = (N_I)_\varphi$. Furthermore, the following results prove that it also preserves, in some sense, (LI) and (CP(N)).

Proposition 2. *Let I be an implication, T a t-norm and $\varphi : [0,1] \to [0,1]$ an automorphism. Then*

$$I \text{ satisfies (LI) with } T \Leftrightarrow I_\varphi \text{ satisfies (LI) with } T_\varphi.$$

Proof. Let us suppose that the pair (I, T) satisfies (LI). Then we have

$$\begin{aligned} I_\varphi(T_\varphi(x,y),z) &= \varphi^{-1}(I(T(\varphi(x),\varphi(y)),\varphi(z))) = \varphi^{-1}(I(\varphi(x),I(\varphi(y),\varphi(z)))) \\ &= I_\varphi(x, I_\varphi(y,z)). \end{aligned}$$

So the pair (I_φ, T_φ) satisfies (LI). Reciprocally, we obtain

$$\begin{aligned} I(T(x,y),z) &= \varphi(I_\varphi(T_\varphi(\varphi^{-1}(x),\varphi^{-1}(y)),\varphi^{-1}(z))) \\ &= \varphi(I_\varphi(\varphi^{-1}(x),I_\varphi(\varphi^{-1}(y),\varphi^{-1}(z)))) = I(x,I(y,z)). \end{aligned}$$

Thus the pair (I, T) satisfies (LI). $\square$

Proposition 3. *Let I be an implication, $\varphi : [0,1] \to [0,1]$ an automorphism and N a fuzzy negation. Then*

$$I \text{ satisfies } (CP(N)) \Leftrightarrow I_\varphi \text{ satisfies } (CP(N_\varphi)).$$

Proof. Let us suppose that I satisfies $(CP(N))$. Then we have

$$
\begin{aligned}
I_\varphi(x,y) &= \varphi^{-1}(I(\varphi(x),\varphi(y))) = \varphi^{-1}(I(N(\varphi(y)),N(\varphi(x)))) \\
&= \varphi^{-1}(I(\varphi(\varphi^{-1}(N(\varphi(y)))),\varphi(\varphi^{-1}(N(\varphi(x)))))) \\
&= \varphi^{-1}(I(\varphi(N_\varphi(y)),\varphi(N_\varphi(x)))) = I_\varphi(N_\varphi(y),N_\varphi(x)).
\end{aligned}
$$

So I_φ satisfies $(CP(N_\varphi))$. Reciprocally, we obtain

$$
\begin{aligned}
I(x,y) &= \varphi(\varphi^{-1}(I(\varphi(\varphi^{-1}(x)),\varphi^{-1}(\varphi(y))))) = \varphi(I_\varphi(\varphi^{-1}(x),\varphi^{-1}(y))) \\
&= \varphi(I_\varphi(N_\varphi(\varphi^{-1}(y)),N_\varphi(\varphi^{-1}(x)))) = \varphi(I_\varphi(\varphi^{-1}(N(y)),\varphi^{-1}(N(x)))) \\
&= I(N(y),N(x)).
\end{aligned}
$$

Thus I satisfies $(CP(N))$. $\qquad\square$

The preservation of all these properties implies that the φ-conjugate of a (S,N)-implication $I_{S,N}$ is the (S,N)-implication I_{S_φ,N_φ}, the φ-conjugate of an R-implication I_T is the R-implication I_{T_φ}, the φ-conjugate of a QL-operation $I_{T,S,N}$ is the QL-operation $I_{T_\varphi,S_\varphi,N_\varphi}$ (see Theorem 2.6.11 in [5]) and finally, a similar fact happens for D-operations since

$$
\begin{aligned}
(I^{T,S,N})_\varphi(x,y) &= \varphi^{-1}(I^{T,S,N}(\varphi(x),\varphi(y))) = \varphi^{-1}(S(T(N(\varphi(x)),N(\varphi(y))),\varphi(y))) \\
&= S_\varphi(\varphi^{-1}(T(N(\varphi(x)),N(\varphi(y))),y)) \\
&= S_\varphi(T_\varphi(\varphi^{-1}(N(\varphi(x))),\varphi^{-1}(N(\varphi(y)))))) \\
&= S_\varphi(T_\varphi(N_\varphi(x),N_\varphi(y)),y) \\
&= I^{T_\varphi,S_\varphi,N_\varphi}(x,y).
\end{aligned}
$$

At this point, let us recall the properties preserved by the min or max-methods. While both methods preserve (NP), (IP) and (OP), they do not preserve either (EP) or (LI). Now the following result shows that they also preserve $(CP(N))$.

Proposition 4. *Let I and J be two implications and N a fuzzy negation. If I and J satisfy $(CP(N))$, $I \vee J$ and $I \wedge J$ satisfy $(CP(N))$.*

Proof. If I and J are two implications satisfying $(CP(N))$, we obtain

$$
\begin{aligned}
(I \vee J)(x,y) &= \max\{I(x,y),J(x,y)\} = \max\{I(N(y),N(x)),J(N(y),N(x))\} \\
&= (I \vee J)(N(y),N(x)).
\end{aligned}
$$

So, $(I \vee J)$ satisfies $(CP(N))$. Analogously, it can be proved for the min-method. $\quad\square$

Finally, the natural negation of the implications generated by these two methods can be easily computed.

Proposition 5. *Let I and J be two implications. Then the natural negations of $I \vee J$ and $I \wedge J$ are*

$$N_{I \vee J}(x) = \max\{N_I(x), N_J(x)\}, \quad N_{I \wedge J}(x) = \min\{N_I(x), N_J(x)\}.$$

Moreover,

(i) if N_I and N_J are continuous (strict), $N_{I \vee J}$ and $N_{I \wedge J}$ are continuous (strict),
(ii) if N_I and N_J are strong, $N_{I \vee J}$ and $N_{I \wedge J}$ are strong.

Proof. We will prove the results only for $I \vee J$. We have that

$$N_{I \vee J}(x) = (I \vee J)(x, 0) = \max\{I(x, 0), J(x, 0)\} = \max\{N_I(x), N_J(x)\}.$$

Next, (i) is due to the fact that the maximum of continuous (strict) functions is a continuous (strict) function. Now, let us prove (ii). Suppose that N_I and N_J are strong. If $N_I(x) \leq N_J(x)$, then we have $N_{I \vee J}(x) = N_J(x)$ and $x \geq N_I(N_J(x))$. Thus we obtain

$$N_{I \vee J}(N_{I \vee J}(x)) = N_{I \vee J}(N_J(x)) = \max\{N_I(N_J(x)), N_J(N_J(x))\} = \max\{N_I(N_J(x)), x\}$$
$$= x.$$

The other case is analogous and consequently, $N_{I \vee J}$ is strong.   □

Remark 1. In some cases, $N_{I \vee J}$ can be strong although N_I and N_J are non-continuous negations. Let us consider

$$N_I(x) = \begin{cases} 1 - x & \text{if } x \leq \frac{1}{2}, \\ 0 & \text{otherwise,} \end{cases}, \quad N_J(x) = \begin{cases} 1 & \text{if } x = 0, \\ \frac{1}{2} & \text{if } 0 < x \leq \frac{1}{2}, \\ 1 - x & \text{otherwise.} \end{cases}$$

Then, $N_{I \vee J}(x) = 1 - x = N_C(x)$ and $N_{I \vee J}$ is a strong negation.

The last method considered in this section is the convex combination of fuzzy implications. As it happens with the previous method, the convex combination preserves (NP), (OP) and (IP), but it does not preserve either (EP) or (LI). On the other hand, it also preserves (CP(N)) and the natural negation of $I_{I,J}^\lambda$ can be expressed in terms of the natural negations of I and J.

Proposition 6. *Let I and J be two implications, $\lambda \in [0, 1]$ and N a fuzzy negation. If I and J satisfy (CP(N)), $I_{I,J}^\lambda$ satisfies (CP(N)).*

Proof. If I and J satisfy (CP(N)), then we obtain

$$I_{I,J}^\lambda(x, y) = \lambda I(x, y) + (1 - \lambda)J(x, y) = \lambda I(N(y), N(x)) + (1 - \lambda)J(N(y), N(x))$$
$$= I_{I,J}^\lambda(N(y), N(x)).$$

So, $I_{I,J}^\lambda$ satisfies (CP(N)).   □

Proposition 7. *Let I and J be two implications and $\lambda \in [0,1]$. The natural negation of $I_{I,J}^{\lambda}$ is*

$$N_{I_{I,J}^{\lambda}}(x) = \lambda N_I(x) + (1-\lambda)N_J(x).$$

Moreover, if N_I and N_J are continuous (strict), $N_{I_{I,J}^{\lambda}}$ is continuous (strict).

Proof. The expression of the natural negation comes from

$$N_{I_{I,J}^{\lambda}}(x) = I_{I,J}^{\lambda}(x,0) = \lambda I(x,0) + (1-\lambda)J(x,0) = \lambda N_I(x) + (1-\lambda)N_J(x).$$

The rest of the result follows due to the fact that a convex combination of continuous (strict) functions is a continuous (strict) function. □

Remark 2. Note that if N_I and N_J are strong negations, $N_{I_{I,J}^{\lambda}}$ could be not strong. Consider $N_I = N_C$, $N_J(x) = \sqrt{1-x^2}$ and $\lambda = \frac{1}{2}$. Then we have

$$N_{I_{I,J}^{\lambda}}(x) = \frac{1}{2} - \frac{x}{2} + \frac{\sqrt{1-x^2}}{2},$$

and taking $x = 0.5$, we obtain

$$N_{I_{I,J}^{\lambda}}(N_{I_{I,J}^{\lambda}}(0.5)) = N_{I_{I,J}^{\lambda}}(0.683013) = 0.523697 \neq 0.5.$$

4 New Construction Methods

In this section, the main goal is to collect the new construction methods presented in the last years and to study which additional properties are satisfied by these new models. The section is divided according to the strategy and initial functions used to generate the new implications. We omit the new classes of implications generated from additive generators, notably studied in the other chapters [38, 20] of this book.

4.1 *Implications Generated from Aggregation Functions*

As it is stated in the introduction, the usual way to generate fuzzy implications is through adequate combinations of aggregation functions and/or fuzzy negations. This strategy is still very active and some new classes of implications have been introduced. Initially, only t-norms and t-conorms were used to derive fuzzy implications. A first generalization was done with the use of conjunctive and disjunctive uninorms for this purpose (main results on this topic were collected in the book [5]). Residual implications have been also derived from conjunctors in general (see [16]) including quasi-copulas and copulas. In such paper some characterizations and construction methods of R-implications are analysed that will not be included in the current survey.

The first two classes we are going to study are the new generalizations of (S,N)-implications taking a different class of aggregation function instead of the usual t-conorm S. The first one is the so-called class of (TS,N)-implications, which are material implications generated from a TS-function and a fuzzy negation.

Definition 6. *(Definition 5 in [11])* A bivariate TS function is a function $TS_{\lambda,f} :$ $[0,1]^2 \rightarrow [0,1]$ defined as

$$TS_{\lambda,f}(x,y) = f^{-1}((1-\lambda)f(T(x,y)) + \lambda f(S(x,y)))$$

for any $x,y \in [0,1]$, where T is a t-norm, S is a t-conorm, $\lambda \in (0,1)$ and $f : [0,1] \rightarrow [-\infty,+\infty]$ is a continuous and strictly monotone function.

Definition 7. *(Definition 10 in [11])* Let $TS_{\lambda,f}$ be a TS-function and let N be a negation. The function $I_{TS,N} : [0,1]^2 \rightarrow [0,1]$ defined as

$$I_{TS,N}(x,y) = TS_{\lambda,f}(N(x),y), \quad x,y \in [0,1],$$

will be called the (TS,N)-operation derived from $TS_{\lambda,f}$ and N.

As QL and D-operations, (TS,N)-operations are not implications in general in the sense of Definition 4 since $I(0,0) = I(1,1) = 1$ can fail. However, the following result shows necessary and sufficient conditions to guarantee that a (TS,N)-operation is in fact a (TS,N)-implication.

Proposition 8. (Proposition 14 in [11]) *Let $TS_{\lambda,f}$ be a TS-function and let N be a negation. The (TS,N)-operation derived from $TS_{\lambda,f}$ and N is a (TS,N)-implication if, and only if, $f(1) = \pm\infty$ and ($f(0) \neq \pm\infty$ or $f(0) + f(1) = f(1)$).*

Example 3. Choosing $f(x) = \frac{1}{1-x}$, the following (TS,N)-implication is obtained

$$I_{TS,N}(x,y) = \frac{\lambda\left(S(N(x),y) - T(N(x),y)\right) + T(N(x),y) - S(N(x),y)T(N(x),y)}{1 - S(N(x),y) + \lambda\left(S(N(x),y) - T(N(x),y)\right)}.$$

In Figure 2, some examples of this family of implications taking $N = N_C$, $\lambda = 0.5$ and first, $T = T_M$ and $S = S_M$, and then $T = T_{LK}$ and $S = S_{LK}$ are displayed.

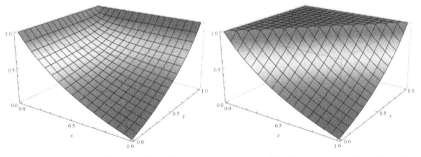

   (a) $I_{TS,N}$ with $T = T_M$, $S = S_M$, $N = N_C$ (b) $I_{TS,N}$ with $T = T_{LK}$, $S = S_{LK}$, $N = N_C$

Fig. 2 Plots of some (TS,N)-implications

This class of fuzzy implications has interesting properties, collected in the next result.

Proposition 9. (Proposition 20 in [11]) *Let $I_{TS,N}$ be a (TS,N)-implication.*

(i) $I_{TS,N}$ satisfies (IP) if, and only if, $S(N(x),x) = 1$ for all $x \in [0,1]$.
(ii) $I_{TS,N}$ does not satisfy (NP).
(iii) If N is strong, then $I_{TS,N}$ satisfies (CP(N)).
(iv) $I_{TS,N}$ satisfies (OP) if, and only if, $N = N_S$ is strong, where $N_S(x) = \inf\{t \in [0,1] : S(x,t) = 1\}$, and $S(N(x),x) = 1$ for all $x \in [0,1]$.

Remark 3. This class of implications does not satisfy (EP) in general. For example, the first implication of Figure 2 does not satisfy (EP) since taking $x = 0.5$, $y = 0.25$ and $z = 0.5$ we obtain

$$I_{TS,N}(0.5, I_{TS,N}(0.25, 0.5)) = I_{TS,N}(0.5, 0.667) = 0.6 \neq 0.667 = I_{TS,N}(0.25, 0.5)$$
$$= I_{TS,N}(0.25, I_{TS,N}(0.5, 0.5)).$$

The other generalization of (S,N)-implications is based on the use of the class of dual representable aggregation functions.

Definition 8. *(Definition 4 in [1])* A binary function $G : [0,1]^2 \to [0,1]$ will be called a dual representable aggregation function (DRAF) if there is a continuous strictly decreasing function $f : [0,1] \to [0+\infty]$ with $f(1) = 0$ and a strong negation N such that G is given by

$$G_{f,N}(x,y) = f^{-1}(\max\{0, f(x \vee y) - f(N(x \wedge y))\}), \quad \text{for all } x,y \in [0,1].$$

Definition 9. *([1])* Let N be a strong negation and let $G_{f,N}$ be a DRAF. The function $I_{G_{f,N},N} : [0,1]^2 \to [0,1]$ defined as

$$I_{G_{f,N},N}(x,y) = G_{f,N}(N(x),y), \quad x,y[0,1],$$

will be called the $(G_{f,N},N)$-implication derived from $G_{f,N}$ and N.

Note that in the previous definition it is considered the same strong negation N that generates the DRAF $G_{f,N}$. There is no impediment to consider other fuzzy negations N' (not necessarily strong) different from N. This could be a future research line.
  The explicit expression of these implications is the following one

$$I_{G_{f,N},N}(x,y) = \begin{cases} 1 & \text{if } x \leq y, \\ f^{-1}(f(y) - f(x)) & \text{if } x > y \geq N(x), \\ f^{-1}(f(N(x)) - f(N(y))) & \text{if } \min\{x, N(x)\} > y. \end{cases}$$

Example 4. Let us show some examples of $(G_{f,N},N)$-implications taking some particular f and N. Some of these implications are showed in Figure 3.

(i) If we take $f(x) = 1 - x$ and $N = N_C$, we obtain the Łukasiewicz implication I_{LK}.

(ii) If we consider $f(x) = \cos\left(\frac{\pi}{2}x\right)$ and $N(x) = \sqrt{1-x^2}$, we get the following implication $I_{G_{f,N},N}(x,y) =$

$$= \begin{cases} 1 & \text{if } x \leq y, \\ \frac{2}{\pi}\arccos\left(\cos\left(\frac{\pi}{2}y\right) - \cos\left(\frac{\pi}{2}x\right)\right) & \text{if } x > y \geq \sqrt{1-x^2}, \\ \frac{2}{\pi}\arccos\left(\cos\left(\frac{\pi}{2}\sqrt{1-x^2}\right) - \cos\left(\frac{\pi}{2}\sqrt{1-y^2}\right)\right) & \text{if } \min\{x, \sqrt{1-x^2}\} > y. \end{cases}$$

(iii) If we consider $f(x) = -\ln(x)$ and $N = N_C$, we get the following implication

$$I_{G_{f,N},N}(x,y) = \begin{cases} 1 & \text{if } x \leq y, \\ \frac{y}{x} & \text{if } x > y \geq 1 - x, \\ \frac{1-x}{1-y} & \text{if } \min\{x, 1-x\} > y. \end{cases}$$

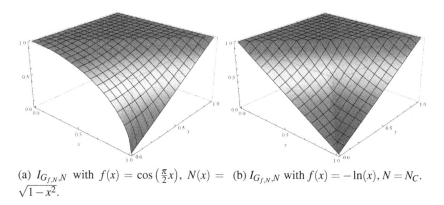

(a) $I_{G_{f,N},N}$ with $f(x) = \cos\left(\frac{\pi}{2}x\right)$, $N(x) = \sqrt{1-x^2}$.

(b) $I_{G_{f,N},N}$ with $f(x) = -\ln(x)$, $N = N_C$.

Fig. 3 Plots of some $(G_{f,N}, N)$-implications

The material implications derived from DRAFs satisfy interesting properties, collected in the next result.

Proposition 10. (Propositions 8 and 9 in ([1])) *Let $G_{f,N}$ be a DRAF and $I_{G_{f,N},N}$ its derived $(G_{f,N}, N)$-implication. Then the following properties hold:*

(i) $I_{G_{f,N},N}$ *satisfies (NP).*

(ii) $N_{I_{G_{f,N},N}} = N$.

(iii) $I_{G_{f,N},N}$ *satisfies (OP).*

(iv) $I_{G_{f,N},N}$ *is continuous.*

(v) $I_{G_{f,N},N}$ *satisfies (CP(N')) for some fuzzy negation N' if, and only if, $N = N'$.*

At this point, let us study (EP) for this class of implications.

Proposition 11. *Let $G_{f,N}$ be a DRAF and $I_{G_{f,N},N}$ its derived $(G_{f,N}, N)$-implication. Then $I_{G_{f,N},N}$ satisfies (EP) if, and only if, $G_{f,N}$ is the nilpotent t-conorm with additive generator $f \circ N$ where $f(0) < +\infty$ and $N(x) = f^{-1}(f(0) - f(x))$ for all $x \in [0,1]$.*

Proof. If $I_{G_{f,N},N}$ satisfies (EP), then it holds

$$G_{f,N}(N(x), G_{f,N}(N(y), z)) = G_{f,N}(N(y), G_{f,N}(N(x), z))$$

for all $x, y, z \in [0,1]$. Now using the commutativity of $G_{f,N}$, we obtain

$$G_{f,N}(G_{f,N}(x,y), z) = G_{f,N}(z, G_{f,N}(x,y)) = G_{f,N}(x, G_{f,N}(z,y)) = G_{f,N}(x, G_{f,N}(y,z)).$$

Consequently, $G_{f,N}$ is associative and by Proposition 6 in [1], the result follows. Reciprocally, if $G_{f,N}$ is a nilpotent t-conorm, then $I_{G_{f,N},N}$ is in fact a (S,N)-implication and it satisfies (EP).   □

Proposition 12. *Let $G_{f,N}$ be a DRAF and $I_{G_{f,N},N}$ its derived $(G_{f,N}, N)$-implication. Then*

(i) *$I_{G_{f,N},N}$ satisfies (LI) if, and only if, $G_{f,N}$ is the nilpotent t-conorm with additive generator $f \circ N$ where $f(0) < +\infty$ and $N(x) = f^{-1}(f(0) - f(x))$ for all $x \in [0,1]$.*
(ii) *$I_{G_{f,N},N}$ satisfies (EP) if, and only if, $I_{G_{f,N},N}$ satisfies (LI).*

Proof. Let us prove (i). If $I_{G_{f,N},N}$ satisfies (LI), then $I_{G_{f,N},N}$ satisfies (EP) and using Proposition 11, the result follows. Reciprocally, if $G_{f,N}$ is a nilpotent t-conorm, $I_{G_{f,N},N}$ is in fact a (S,N)-implication derived from a strong negation and consequently, it satisfies (LI). Finally, (ii) is straightforward.   □

Another new class of fuzzy implications is the generalization of R-implications choosing as conjunctions the so-called representable aggregation functions.

Definition 10. *(Definition 2.4 in [14])* A binary function $F : [0,1]^2 \to [0,1]$ will be called a representable aggregation function (RAF) if there is a continuous strictly increasing function $g : [0,1] \to [0, +\infty]$ with $g(0) = 0$ and a strong negation N such that F is given by

$$F_{g,N}(x,y) = g^{-1}(\max\{0, g(x \wedge y) - g(N(x \vee y))\}), \quad \text{for all } x, y \in [0,1].$$

Definition 11. *([14])* Let $F_{g,N}$ be a RAF. The function $I_{F_{g,N}} : [0,1]^2 \to [0,1]$ defined as

$$I_{F_{g,N}}(x,y) = \sup\{z \in [0,1] | F(x,z) \le y\}, \quad x, y \in [0,1],$$

will be called the R-implication derived from $F_{g,N}$.

The election of a strong negation into the definitions of RAFs and DRAFs ensures the duality between these two classes of aggregation functions, but RAFs or DRAFs generated from non-strong negations also generate residual or material implications, respectively, satisfying the conditions of Definition 4.

As we have already said, it is straightforward that the previous definition always gives an implication in the sense of Definition 4 since RAFs are conjunctors. A more detailed expression of this class of implications is the following one

$$I_{F_{g,N}}(x,y) = \begin{cases} 1 & \text{if } x \leq y, \\ g^{-1}(g(N(x)) + g(y)) & \text{if } x > s \text{ and } y < \delta(x), \\ N(g^{-1}(g(x) - g(y))) & \text{otherwise,} \end{cases}$$

where s is the fixed point of N and δ is the diagonal section of $F_{g,N}$, that is $\delta(x) = F_{g,N}(x,x)$ for all $x \in [0,1]$.

Example 5. Let us show some examples of R-implications derived from RAFs taking some particular g and N. Some of these implications are plotted in Figure 4.

(i) If we consider $g = id_{[0,1]}$ and $N = N_C$, we obtain the usual Łukasiewicz implication I_{LK}.

(ii) If we consider $g(x) = x^2$ and $N = N_C$, we get the following implication

$$I_{F_{g,N}}(x,y) = \begin{cases} 1 & \text{if } x \leq y, \\ \sqrt{1 + x^2 + y^2 - 2x} & \text{if } x > \frac{1}{2} \text{ and } y < \sqrt{\max\{0, 2x-1\}}, \\ 1 - \sqrt{x^2 - y^2} & \text{otherwise.} \end{cases}$$

(iii) If we consider $g(x) = x^2$ and $N(x) = \sqrt{1-x^2}$, we get the following implication

$$I_{F_{g,N}}(x,y) = \begin{cases} 1 & \text{if } x \leq y, \\ \sqrt{1 + y^2 - x^2} & \text{otherwise.} \end{cases}$$

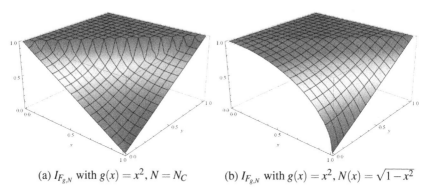

    (a) $I_{F_{g,N}}$ with $g(x) = x^2, N = N_C$ (b) $I_{F_{g,N}}$ with $g(x) = x^2, N(x) = \sqrt{1-x^2}$

Fig. 4 Plots of some R-implications from RAFs

Now, let us study which additional properties are satisfied by these implications. Note that the fulfilment of some important properties such as (EP) or (CP(N)) is directly connected to consider as a RAF the φ-conjugate of the Łukasiewicz t-norm.

Proposition 13. Proposition 3.3 and Corollary 3.7 in ([14]) *Let $F_{g,N}$ be a RAF and $I_{F_{g,N}}$ be the R-implication derived from $F_{g,N}$. Then the following properties hold:*

(i) $N_{I_{F_{g,N}}} = N$.
(ii) $I_{F_{g,N}}$ *satisfies (NP)*.
(iii) $I_{F_{g,N}}$ *satisfies (OP)*.
(iv) $I_{F_{g,N}}$ *is continuous*.
(v) $I_{F_{g,N}}$ *satisfies (EP) or (CP(N)) if, and only if, there exists some automorphism φ such that $F_{g,N} = (T_{LK})_{\varphi}$*.

From the last point of the previous result, the law of importation for this class of implications can be solved.

Proposition 14. *Let $F_{g,N}$ be a RAF and $I_{F_{g,N}}$ be the R-implication derived from $F_{g,N}$. Then*

(i) $I_{F_{g,N}}$ *satisfies (LI) if, and only if, there exists some automorphism φ such that $F_{g,N} = (T_{LK})_{\varphi}$*.
(ii) $I_{F_{g,N}}$ *satisfies (EP) if, and only if, $I_{F_{g,N}}$ satisfies (LI)*.

Proof. Let us prove (i). If $I_{F_{g,N}}$ satisfies (LI), then $I_{F_{g,N}}$ satisfies (EP) and using point (v) in Proposition 13, $F_{g,N} = (T_{LK})_{\varphi}$. Reciprocally, if $F_{g,N} = (T_{LK})_{\varphi}$, $I_{F_{g,N}}$ is in fact an R-implication derived from a t-norm and consequently, it satisfies (LI). Finally, (ii) is straightforward. □

Finally, just to mention that a more general attempt to generalize (S,N)-implications with a strong negation and R-implications to other aggregation functions is made in [35]. In that paper, the definitions of R-implication and S-implication like operators are given from a general binary operator $A : [0,1]^2 \to [0,1]$ and a strong negation N respectively in the following way

$$I_A(x,y) = \sup\{z | A(x,z) \le y\}, \qquad x,y \in [0,1].$$
$$I_{A,N}(x,y) = A(N(x),y),$$

These general operators are not implications in general in the sense of Definition 4. After that, some sufficient conditions in order to obtain fuzzy implications from these operators satisfying certain additional properties are stated (see [35]).

4.2 Implications Generated from Negations

From now on, we will deal with implications generated only from fuzzy negations. This approach is quite novel, but in some sense it is related to the generation of fuzzy implications through the use of additive generators. Note that for example an f-generator, i.e., a strictly decreasing function $f : [0,1] \to [0,+\infty]$ such that $f(1) = 0$

that also satisfies $f(0) = 1$, is in fact a strict negation. However although there exists some intersections between the implications generated by the two approaches, they generate different implications in general.

The first method was introduced by Jayaram and Mesiar in [21] while they were studying special implications. A fuzzy implication I is special if for any $\varepsilon > 0$ and for all $x, y \in [0, 1]$ such that $x + \varepsilon, y + \varepsilon \in [0, 1]$ the following condition is satisfied:

$$I(x, y) \leq I(x + \varepsilon, y + \varepsilon). \tag{SP}$$

From this study, they introduced the neutral special implications with a given negation.

Definition 12. *([21])* Let N be a fuzzy negation such that $N \leq N_C$. Then the function $I_N : [0, 1]^2 \to [0, 1]$ given by

$$I_N(x, y) = \begin{cases} 1 & \text{if } x \leq y, \\ y + \frac{N(x-y)(1-x)}{1-x+y} & \text{if } x > y, \end{cases}$$

with the understanding $\frac{0}{0} = 0$, is called the neutral special implication generated from N.

Remark 4. The condition $N \leq N_C$ was imposed to ensure the generation of a special implication satisfying (NP). However, any fuzzy negation N can be used in order to generate this class of implications.

The class of neutral special implications contains some well-known implications as the following example shows.

Example 6. Let us find some examples of neutral special implications taking different fuzzy negations. The plot of some of these implications can be viewed in Figure 5.

(i) If we consider $N = N_C$, we obtain the Łukasiewicz implication I_{LK}.
(ii) If we consider $N = N_{D_1}$, we get the Gödel implication I_{GD}.
(iii) If we take $N(x) = (1 - \sqrt{x})^2$, we obtain the implication given by

$$I_N(x, y) = \begin{cases} 1 & \text{if } x \leq y, \\ y + \frac{(1-\sqrt{x-y})^2(1-x)}{1-x+y} & \text{if } x > y. \end{cases}$$

(iv) If we take the Sugeno negation with $\lambda = 2$, the following implication is obtained

$$I_N(x, y) = \begin{cases} 1 & \text{if } x \leq y, \\ y + \frac{1-x}{1+2x-2y} & \text{if } x > y. \end{cases}$$

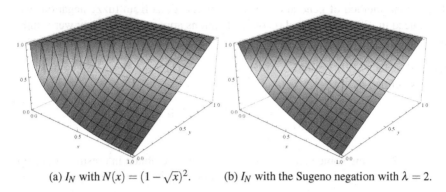

(a) I_N with $N(x) = (1 - \sqrt{x})^2$.   (b) I_N with the Sugeno negation with $\lambda = 2$.

Fig. 5 Plots of some neutral special implications

The next result shows which properties in addition to (NP) and (SP) are satisfied by the neutral special implications.

Proposition 15. ([21]) *Let N be a fuzzy negation such that $N \leq N_C$ and I_N its derived neutral special implication. The following properties hold:*

(i) I_N *satisfies (NP).*
(ii) I_N *satisfies (SP).*
(iii) I_N *satisfies (OP).*
(iv) $N_{I_N} = N$.

Furthermore, some partial results can be proved with respect to the contrapositive symmetry.

Proposition 16. *Let N be a fuzzy negation such that $N \leq N_C$ and I_N its derived neutral special implication. Then if I_N satisfies (CP(N')) with a negation N' then $N' = N$ is a strong negation.*

Proof. If I_N satisfies (CP(N')) with a negation N', since I_N satisfies (NP), using Corollary 1.5.8 in [5] $N' = N$ is a strong negation. □

However, this condition is not enough to ensure the fulfilment of (CP(N_I)) as the implication obtained in Example 6-(iv) shows. Let N be the Sugeno negation with $\lambda = 2$. Consider $x = 0.5$ and $y = 0.3$, then we obtain

$$I_N(0.5, 0.3) = 0.657143 \neq 0.659091 = I_N(0.4375, 0.25) = I_N(N(0.3), N(0.5)).$$

In addition, the same implication shows that I_N do not satisfy (EP) in general. If we take $x = 0.9$, $y = 0.8$ and $z = 0.7$, we have

$$I_N(0.9, I_N(0.8, 0.7)) = I_N(0.9, 0.866667) = 0.960417 \neq 0.960618$$
$$= I_N(0.8, 0.771429) = I_N(0.8, I_N(0.7, 0.9)).$$

The second method of generation of fuzzy implications from fuzzy negations was introduced in [39]. This new class of implications appeared while they were studying the dependencies and independencies of several fuzzy implication properties.

Definition 13. *([39])* Let N be a fuzzy negation. Then the function $I^N : [0,1]^2 \to [0,1]$ is defined by

$$I^N(x,y) = \begin{cases} 1 & \text{if } x \leq y, \\ \frac{(1-N(x))y}{x} + N(x) & \text{if } x > y. \end{cases}$$

Example 7. Taking some concrete fuzzy negations, we obtain interesting fuzzy implications. The plot of some of these implications can be viewed in Figure 6.

(i) If we consider $N = N_C$, we obtain the Łukasiewicz implication I_{LK}.
(ii) If we consider $N = N_{D_1}$, we get the Goguen implication I_{GG}.
(iii) If we take $N = N_{D_2}$, we obtain the Weber implication I_{WB}.
(iv) If we take $N(x) = 1 - x^2$, the following implication is obtained

$$I^N(x,y) = \begin{cases} 1 & \text{if } x \leq y, \\ 1 - x^2 + xy & \text{if } x > y. \end{cases}$$

(v) If we take $N(x) = 1 - \sqrt{x}$, we obtain the following implication

$$I^N(x,y) = \begin{cases} 1 & \text{if } x \leq y, \\ 1 - \frac{(x-y)\sqrt{x}}{x} & \text{if } x > y. \end{cases}$$

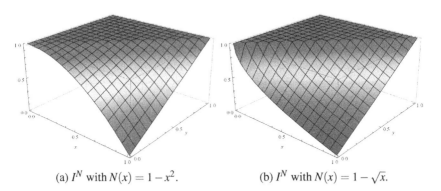

(a) I^N with $N(x) = 1 - x^2$.   (b) I^N with $N(x) = 1 - \sqrt{x}$.

Fig. 6 Plots of some I^N implications

It is clear that this method provides fuzzy implications in the sense of Definition 4. In order to study the additional properties, we need to introduce some generalizations of the negation N_{D_1} and the Sugeno class of fuzzy negations, which are given by

$$N_A(x) = \begin{cases} 1 \text{ if } x \in A, \\ 0 \text{ if } x \notin A, \end{cases}, \quad N_{A,\beta}(x) = \begin{cases} 1 & \text{if } x \in A, \\ \frac{1-x}{1+\beta x} & \text{if } x \notin A, \end{cases}$$

where $A = [0, \alpha)$ with $\alpha \in (0, 1)$ or $A = [0, \alpha]$ with $\alpha \in [0, 1]$. Notice that $N_{\{0\}} = N_{D_1}$ and $N_{\{0\},\beta}$ is the Sugeno class of negations.

Proposition 17. ([39]) *Let N be a fuzzy negation and consider the implication I^N. Then the following properties hold:*

(i) I^N satisfies (NP).
(ii) I^N satisfies (EP) if, and only if, $N = N_A$ or $N = N_{A,\beta}$.
(iii) I^N satisfies (OP) if, and only if, $N(x) = 1 \Leftrightarrow x = 0$, that is, N is a non-filling negation.
(iv) $N_{I^N} = N$.
(v) I^N satisfies (IP).
(vi) I^N satisfies (CP(N')) with a strong fuzzy negation if, and only if, $N = N'$ is a Sugeno negation.

From the previous result, an interesting consequence related to the dependencies of these properties emerges.

Corollary 1. (Corollary 1 in [39]) *Let N be a fuzzy negation and consider the implication I^N. Then the following statements are equivalent:*

(i) N is a Sugeno negation,
(ii) I^N satisfies (EP) and N is a continuous negation,
(iii) I^N satisfies (CP(N)),
(iv) I^N is conjugate with the Łukasiewicz implication, that is, $I^N = (I_{LK})_\varphi$ for some automorphism φ.

To end this part, it is worth to remind that a simple method to generate fuzzy implications only from negations is to fix a concrete aggregation function on the methods based on the use of an aggregation function and a fuzzy negation. For example, if we take $f(x) = 1 - x$ then the associated RAF, $G_{f,N}(x,y) = 1 - \max\{0, N(x \wedge y) - (x \vee y)\}$, where N is any fuzzy negation generates the following family of implications depending on N (see [14]):

$$I_{G_{f,N}}(x,y) = \begin{cases} 1 & \text{if } x \leq y, \\ 1 + y - x & \text{if } x > y \geq N(x), \\ 1 - N(y) + N(x) & \text{if } \min\{x, N(x)\} > y. \end{cases}$$

4.3 Implications Constructed from a Given Implication

Till the end of the paper, we will deal with methods based on the use of at least a fuzzy implication to generate a new one. First of all, we present some generation methods which use a given fuzzy implication and a fuzzy negation to generate a new fuzzy implication. The first methods were introduced when studying which properties are preserved by the threshold horizontal and vertical methods of construction of a fuzzy implication from two given ones (see Section 4.4). Consider a fuzzy implication I, a fuzzy negation N, $e \in (0,1)$, then the following four implications can be generated.

- If $N(x) \in [0,e]$ for all $e < x$,

$$
I_{I,N}^{1v}(x,y) = \begin{cases} 1 & \text{if } x \le e, \\ I\left(N\left(\frac{N(x)}{e}\right),y\right) & \text{if } e < x, \end{cases}
$$

$$
I_{I,N}^{0h}(x,y) = \begin{cases} 1 & \text{if } x = 0, \\ 0 & \text{if } x > 0, y \le e, \\ I\left(x,N\left(\frac{N(y)}{e}\right)\right) & \text{if } x > 0, y > e. \end{cases}
$$

- If $N(x) \in [e,1]$ for all $x < e$,

$$
I_{I,N}^{0v}(x,y) = \begin{cases} 1 & \text{if } y = 1, \\ 0 & \text{if } x \ge e \text{ and } y < 1, \\ I\left(N\left(\frac{N(x)-e}{1-e}\right),y\right) & \text{if } x < e, \end{cases}
$$

$$
I_{I,N}^{1h}(x,y) = \begin{cases} 1 & \text{if } y \ge e, \\ I\left(x,N\left(\frac{N(y)-e}{1-e}\right)\right) & \text{if } y < e. \end{cases}
$$

The plot of these constructions taking $I = I_{LK}$ and $N = N_C$ can be viewed in Figure 7. Note that all these constructions have a flat zero or one zone and in the other region, some scaling on the initial implication is performed through the negation N. The properties preserved by these implications have not been deeply studied, however an interesting result can be formulated on the contrapositive symmetry involving some kind of preservation of (CP(N)) scaled by $e \in (0,1)$.

Proposition 18. (Lemma 21 in [34] and Lemma 3 in [29]) *Let I be a fuzzy implication and N a strong negation with $N(e) = e$. Then*

(i) *I satisfies (CP(N)) $\Leftrightarrow I_{I,N}^{1v}\left(x,\frac{y}{e}\right) = I_{I,N}^{1v}\left(N(y),\frac{N(x)}{e}\right)$ for all $e < x \le 1$ and $0 \le y \le e$.*

(ii) *I satisfies (CP(N)) $\Leftrightarrow I_{I,N}^{0v}\left(x,\frac{y-e}{1-e}\right) = I_{I,N}^{0v}\left(N(y),\frac{N(x)-e}{1-e}\right)$ for all $0 \le x < e$ and $e \le y \le 1$.*

(iii) *If I satisfies (CP(N)) then $I_{I,N}^{0h}\left(\frac{x}{e},y\right) = I_{I,N}^{0h}\left(\frac{N(y)}{e},N(x)\right)$ for all $0 < x < e$ and $e < y < 1$. Moreover, if $I(x,0) = I(1,N(x))$ for all $x \in [0,1]$, the reciprocal holds too.*

(iv) *I satisfies (CP(N)) if and only if $I_{I,N}^{1h}\left(\frac{x-e}{1-e},y\right) = I_{I,N}^{1h}\left(\frac{N(y)-e}{1-e},N(x)\right)$ for all $x \ge e$ and $y \le e$.*

A quite similar property for (EP) is available for the implications generated from an initial implication through the following methods introduced also in [29]:

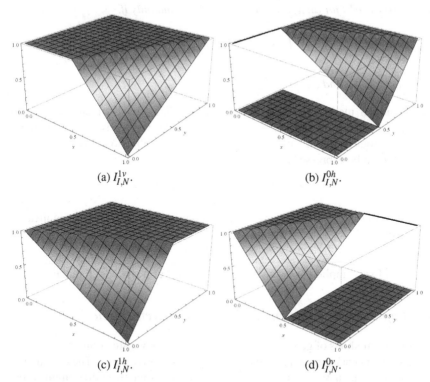

(a) $I_{I,N}^{1v}$.   (b) $I_{I,N}^{0h}$.

(c) $I_{I,N}^{1h}$.   (d) $I_{I,N}^{0v}$.

Fig. 7 Plots of some construction methods from $I = I_{LK}$, $N = N_C$ and $e = \frac{1}{2}$

$$
I_I^{0h}(x,y) = \begin{cases} 1 & \text{if } x = 0, \\ 0 & \text{if } x > 0 \text{ and } y \le e, \\ I\left(ex, \frac{y-e}{1-e}\right) & \text{if } x > 0 \text{ and } y > e, \end{cases}
$$

$$
I_I^{1h}(x,y) = \begin{cases} 1 & \text{if } x = 0, \\ I\left(e + (1-e)x, \frac{y}{e}\right) & \text{if } x > 0 \text{ and } y \le e, \\ 1 & \text{if } x > 0 \text{ and } y > e. \end{cases}
$$

Notice that these implications also have a flat zero or one zone and some scaling in the other region. With these implications, we obtain the following property related to Proposition 18 but now for (EP).

Proposition 19. (Propositions 8 and 9 in [29]) *Let I be a fuzzy implication. Then the following properties hold:*

(i) If $I(x,y) > 0$ when $y > 0$, then I satisfies (EP) for all $x,y < e$ and $z \in (0,1]$ if, and only if,

$$
I_I^{0h}\left(\frac{x}{e}, e + (1-e)I_I^{0h}\left(\frac{y}{e}, z\right)\right) = I_I^{0h}\left(\frac{y}{e}, e + (1-e)I_I^{0h}\left(\frac{x}{e}, z\right)\right)
$$

for all $x,y < e$ and $z \in [0,1]$.

(ii) I satisfies (EP) for all $x, y > e$ and $z \in [0,1]$ if, and only if,

$$I_I^{1h}\left(\frac{x-e}{1-e}, eI_I^{1h}\left(\frac{y-e}{1-e}, z\right)\right) = I_I^{1h}\left(\frac{y-e}{1-e}, eI_I^{1h}\left(\frac{x-e}{1-e}, z\right)\right)$$

for all $x, y \geq e$ and $z \in [0,1]$.

Finally, another method to obtain a fuzzy implication from a given one, a fuzzy negation and a t-conorm has been proposed in [39]. This method is based on noticing that the previous method I^N of generation of a fuzzy implication only from a fuzzy negation N can be expressed as

$$I^N(x, y) = S_P(N(x), I_{GG}(x, y)).$$

From this observation, replacing S_P for any t-conorm S and I_{GG} for any implication I, the function

$$I^{N,S,I}(x, y) = S(N(x), I(x, y)), \quad x, y \in [0,1],$$

is always a fuzzy implication. This topic is worth of further research in the future.

4.4 Implications Constructed from Two Given Implications

The last new methods of generation of fuzzy implication we are going to present are those based on generating a fuzzy implication from two given ones. The first method is based on an adequate scaling of the second variable of the two initial implications and it is called threshold generation method of a fuzzy implication [34].

Definition 14. *(Theorem 3 in [34])* Let I_1 and I_2 be two fuzzy implications and $e \in (0,1)$. The function $I_{I_1-I_2} : [0,1]^2 \to [0,1]$ defined by

$$I_{I_1-I_2}(x, y) = \begin{cases} 1 & \text{if } x = 0, \\ e \cdot I_1\left(x, \frac{y}{e}\right) & \text{if } x > 0 \text{ and } y \leq e, \\ e + (1-e) \cdot I_2\left(x, \frac{y-e}{1-e}\right) & \text{if } x > 0 \text{ and } y > e, \end{cases}$$

is called the e-threshold generated implication from I_1 and I_2.

Example 8. If we consider the Łukasiewicz implication and the Fodor implication, the threshold generated implication from these two implications is given by

$$I_{I_{LK}-I_{FD}}(x, y) = \begin{cases} 1 & \text{if } x = 0, \\ \min\{e, e - ex + y\} & \text{if } x > 0 \text{ and } y \leq e, \\ \max\{1 - x + ex, y\} & \text{if } x > 0, y > e \text{ and } x > \frac{y-e}{1-e}, \\ 1 & \text{otherwise.} \end{cases}$$

In Figure 8, the effect of the choice of the e threshold value is clearly visible.

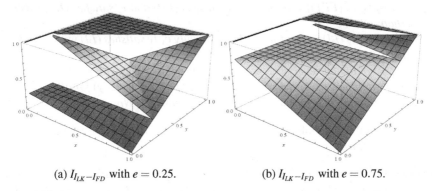

(a) $I_{I_{LK}-I_{FD}}$ with $e = 0.25$.   (b) $I_{I_{LK}-I_{FD}}$ with $e = 0.75$.

Fig. 8 Plots of some e-threshold generated implications with different values of e

This method always generates fuzzy implications and it allows a certain degree of control of the increasingness in the second variable of the values of the implication as the following result shows.

Proposition 20. (Proposition 4 in [34]) *Let I_1 and I_2 be two implications. Then*

(i) $I_{I_1-I_2}(x,y) \leq e$ *if $x > 0$ and $y < e$,*
$I_{I_1-I_2}(x,e) = e$ *if $x > 0$,*
$I_{I_1-I_2}(x,y) \geq e$ *if $x > 0$ and $y > e$.*
(ii) *When $x > 0$ and $y < e$ then $I_{I_1-I_2}(x,y) < e \Leftrightarrow I_1(a,b) < 1$ for $a > 0$, $b < 1$.*
(iii) *When $x > 0$ and $y > e$ then $I_{I_1-I_2}(x,y) > e \Leftrightarrow I_2(a,b) > 0$ for $a,b > 0$.*

Furthermore, the importance of this method derives from the fact that it allows to characterize h-implications [28] since any h-implication is the threshold generated implication of a Yager's f-generated and a Yager's g-generated implication (see Theorem 2 and Remark 30 in [34]). In addition, in contrast to other generation methods of fuzzy implications from two given ones, it preserves (EP) and (LI) from the initial implications to the generated one.

Proposition 21. ([31, 34]) *Let I_1, I_2 be two implications. Then the following properties hold:*

(i) *I_1 and I_2 satisfy (NP) if, and only if, $I_{I_1-I_2}$ satisfies (NP).*
(ii) *$I_{I_1-I_2}$ satisfies neither (OP) nor (IP).*
(iii) *The natural negation of a threshold generated implication is*

$$N_{I_{I_1-I_2}}(x) = \begin{cases} 1 & \text{if } x = 0, \\ eN_{I_1}(x) & \text{if } x > 0. \end{cases}$$

(iv) *If N is a strong negation with fixed point $e \in (0,1)$, then e-threshold generated implication $I_{I_1^{1v},N-I_{2,N}^{0v}}$ satisfies (CP(N)) if, and only if, I_1 satisfies (CP(N)) and I_2 satisfies (CP(N)) except perhaps when $y = 0$.*

(v) $I_{I_1-I_2}$ *satisfies (EP) if, and only if, I_1 satisfies (EP) and I_2 satisfies (EP) except perhaps when $z = 0$.*

(vi) $I_{I_1-I_2}$ *satisfies (LI) with a t-norm T if, and only if, I_1 satisfies (LI) with T and I_2 satisfies (LI) with T except perhaps when $z = 0$.*

(vii) $I_{I_1-I_2}$ *is continuous everywhere except at the points $(0,y)$ with $y \le e$ if and only if, I_1 is continuous everywhere except maybe at the points $(0,y)$ for all $y \in [0,1]$, I_2 is continuous everywhere except at $(0,0)$ and $N_{I_2} = N_{D_1}$.*

In addition, all the threshold generated implications can be characterized as the following result shows.

Theorem 1. (Theorem 14 in [34]) *Let I be a fuzzy implication and $e \in (0,1)$. Then I is an e-threshold generated implication $I_{I_1-I_2}$ if, and only if, $I(x,e) = e$ for all $x > 0$. In this case, the initial implications I_1 and I_2 are respectively given by*

$$I_1(x,y) = \begin{cases} \frac{I(x,ey)}{e} & \text{if } x > 0, \\ 1 & \text{if } x = 0, \end{cases} \qquad I_2(x,y) = \frac{I(x,e+(1-e)\cdot y) - e}{1-e}.$$

The previous threshold method is based on splitting the domain of the implication with a horizontal line and then scaling the two initial implications in order to be well-defined in those two regions. An analogous method can be made but now using a vertical line. This is the idea behind the vertical threshold generation method of a fuzzy implication.

Definition 15. *(Theorem 3 in [29])* Let I_1 and I_2 be two fuzzy implications and $e \in (0,1)$. The function $I_{I_1|I_2} : [0,1]^2 \to [0,1]$ defined by

$$I_{I_1|I_2}(x,y) = \begin{cases} e+(1-e)\cdot I_1\left(\frac{x}{e},y\right) & \text{if } x < e, y < 1, \\ e\cdot I_2\left(\frac{x-e}{1-e},y\right) & \text{if } x \ge e, y < 1, \\ 1 & \text{if } y = 1, \end{cases}$$

is called the vertical e-threshold generated implication from I_1 and I_2.

Example 9. If we consider again the Łukasiewicz implication and the Fodor implication, the vertical e-threshold generated implication from these two implications is given by

$$I_{I_{LK}|I_{FD}}(x,y) = \begin{cases} 1 & \text{if } y = 1, \\ \min\{1, 1-\frac{x}{e}+x+y(1-e)\} & \text{if } x < e \text{ and } y < 1, \\ e & \text{if } x \ge e, y < 1 \text{ and } \frac{x-e}{1-e} \le y, \\ \max\{\frac{e(1-x)}{1-e}, ey\} & \text{otherwise.} \end{cases}$$

In Figure 9, it can be seen how the vertical line $x = e$ splits the domain in two zones.

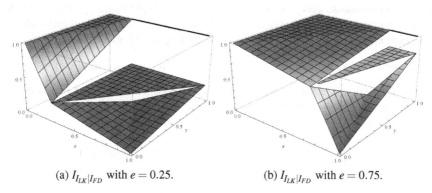

(a) $I_{I_{LK}|I_{FD}}$ with $e = 0.25$.   (b) $I_{I_{LK}|I_{FD}}$ with $e = 0.75$.

Fig. 9 Plots of some vertical e-threshold generated implications with different values of e

As a logical consequence of the definition of the vertical threshold generated implications, the values of the implication are below e when the first variable is over e (except when $y = 1$) and vice-versa.

Proposition 22. (Proposition 1 in [29]) *Let I_1 and I_2 be two implications. Then*

(i) $I_{I_1|I_2}(x,y) \leq e$ if $x > e$ and $y < 1$,
   $I_{I_1|I_2}(e,y) = e$ if $y < 1$,
   $I_{I_1|I_2}(x,y) \geq e$ if $x < e$ and $y < 1$.
(ii) When $x < e$ and $y < 1$ then $I_{I_1|I_2}(x,y) > e \Leftrightarrow I_1(a,b) > 0$ for $a,b < 1$.
(iii) When $x > e$ and $y < 1$ then $I_{I_1|I_2}(x,y) < e \Leftrightarrow I_2(a,b) < 1$ for $a > 0$, $b < 1$.

This method does not preserve so many properties as the horizontal threshold method, but some results can be proved. The following ones were presented in [29] without proofs. An extended version including proofs and additional results is forthcoming [33].

Proposition 23. ([29, 33]) *Let I_1, I_2 be two implications. Then the following properties hold:*

(i) $I_{I_1|I_2}$ does not satisfy either (NP), (IP) or (OP).
(ii) The natural negation of a vertical e-threshold generated implication is

$$N_{I_1|I_2}(x) = \begin{cases} e + (1-e)N_{I_1}\left(\frac{x}{e}\right) & \text{if } x < e, \\ eN_{I_2}\left(\frac{x-e}{1-e}\right) & \text{if } x \geq e. \end{cases}$$

(iii) Suppose I_1 and I_2 such that $I_1(x,y) > 0$ for all $y > 0$ and $I_1(x,y) < 1$ for all $0 < x \leq e$ and $0 < y < 1$. Then $I_{I_1^0|I_2^1}$ satisfies (EP) if, and only if, I_1 satisfies (EP) for all $0 < x,y < e$ and $z \in (0,1)$ and I_2 satisfies (EP) for all $x,y \geq e$ and $z \in [0,1]$.
(iv) Let N be a strong fuzzy negation with fixed point e. Then I_1 satisfies (CP(N)) for all $(x,y) \in (0,1)^2$ and I_2 satisfies (CP(N)) if, and only if, the vertical e-threshold generated implication $I_{I_{1,N}^0|I_{2,N}^1}$ satisfies (CP(N)).

(v) $I_{I_1|I_2}$ *is continuous everywhere except at the points* $(x, 1)$ *with* $x \geq e$ *if and only if* I_1 *is continuous everywhere except maybe at* $(1, 1)$, I_2 *is continuous everywhere except maybe at the points* $(x, 1)$ *with* $x \in [0, 1]$ *and* $I_1(1, y) = 0$ *for all* $y < 1$.

Finally, we can also characterize the vertical threshold generated implications.

Theorem 2. (Theorem 3 in [29]) *Let* I *be a fuzzy implication. Then* I *is a vertical e-threshold generated implication* $I_{I_1|I_2}$ *if, and only if,* $I(e, y) = e$ *for all* $y < 1$. *In this case, the initial implications* I_1 *and* I_2 *are respectively given by*

$$I_1(x, y) = \frac{I(ex, y) - e}{1 - e}, \quad I_2(x, y) = \begin{cases} \frac{I(e + (1-e) \cdot x, y)}{e} & \text{if } y < 1, \\ 1 & \text{if } y = 1. \end{cases}$$

5 Conclusions

In this paper, we have presented the recently introduced methods of generation of fuzzy implications and we have studied the additional properties which are satisfied by these new implications. In addition, we have recalled the usual methods to obtain fuzzy implications analysing some properties which have not been studied yet.

Among the different methods, we have highlighted those methods that generate new implications from old ones. If we restrict the importance of these methods to the preservation of the properties of the initial implication(s), two methods stand out: the conjugation method and the horizontal threshold generation method. These are the only methods which preserve (EP) and even (LI). On the other hand, simple methods like the max or min-methods or the convex combination method deserve a deeper study in order to characterize the generated implications satisfying (EP) or (LI).

Acknowledgements. This paper has been partially supported by the Spanish Grant MTM2009-10320 with FEDER support.

References

1. Aguiló, I., Carbonell, M., Suñer, J., Torrens, J.: Dual Representable Aggregation Functions and Their Derived S-Implications. In: Hüllermeier, E., Kruse, R., Hoffmann, F. (eds.) IPMU 2010. LNCS, vol. 6178, pp. 408–417. Springer, Heidelberg (2010)
2. Aguiló, I., Suñer, J., Torrens, J.: A characterization of residual implications derived from left-continuous uninorms. Information Sciences 180(20), 3992–4005 (2010)
3. Baczyński, M., Drewniak, J.: Conjugacy Classes of Fuzzy Implications. In: Reusch, B. (ed.) Fuzzy Days 1999. LNCS, vol. 1625, pp. 287–298. Springer, Heidelberg (1999)
4. Baczyński, M., Jayaram, B.: On the characterization of (S,N)-implications. Fuzzy Sets and Systems 158, 1713–1727 (2007)
5. Baczyński, M., Jayaram, B.: Fuzzy Implications. STUDFUZZ, vol. 231. Springer, Heidelberg (2008)
6. Baczyński, M., Jayaram, B.: (S,N)- and R-implications: A state-of-the-art survey. Fuzzy Sets and Systems 159, 1836–1859 (2008)

7. Baczyński, M., Jayaram, B.: (U,N)-implications and their characterizations. Fuzzy Sets and Systems 160, 2049–2062 (2009)
8. Balasubramaniam, J.: Contrapositive symmetrisation of fuzzy implications–revisited. Fuzzy Sets and Systems 157(17), 2291–2310 (2006)
9. Balasubramaniam, J.: Yager's new class of implications J_f and some classical tautologies. Information Sciences 177, 930–946 (2007)
10. Bandler, W., Kohout, L.: Semantics of implication operators and fuzzy relational products. Internat. J. Man-Machine Studies 12, 89–116 (1980)
11. Bustince, H., Fernandez, J., Pradera, A., Beliakov, G.: On (TS,N)-fuzzy implications. In: De Baets, B., Mesiar, R., Troiano, L. (eds.) 6th International Summer School on Aggregation Operators (AGOP 2011), Benevento, Italy, pp. 93–98 (2011)
12. Bustince, H., Mohedano, V., Barrenechea, E., Pagola, M.: Definition and construction of fuzzy DI-subsethood measures. Information Sciences 176, 3190–3231 (2006)
13. Bustince, H., Pagola, M., Barrenechea, E.: Construction of fuzzy indices from fuzzy DI-subsethood measures: application to the global comparison of images. Information Sciences 177, 906–929 (2007)
14. Carbonell, M., Torrens, J.: Continuous R-implications generated from representable aggregation functions. Fuzzy Sets and Systems 161, 2276–2289 (2010)
15. De Baets, B., Fodor, J.C.: Residual operators of uninorms. Soft Computing 3, 89–100 (1999)
16. Durante, F., Klement, E., Mesiar, R., Sempi, C.: Conjunctors and their residual implicators: Characterizations and construction methods. Mediterranean Journal of Mathematics 4, 343–356 (2007)
17. Fodor, J.C.: Contrapositive symmetry of fuzzy implications. Fuzzy Sets and Systems 69(2), 141–156 (1995)
18. Fodor, J.C., Roubens, M.: Fuzzy Preference Modelling and Multicriteria Decision Support. Kluwer Academic Publishers, Dordrecht (1994)
19. Grabisch, M., Marichal, J.-L., Mesiar, R., Pap, E.: Aggregation Functions (Encyclopedia of Mathematics and its Applications), 1st edn. Cambridge University Press, New York (2009)
20. Hlinena, D., Kalina, M., Kral, P.: Implication functions generated using functions of one variable. In: Baczynski, M., Beliakov, G., Bustince, H., Pradera, A. (eds.) Advances in Fuzzy Implication Functions. Springer, Berlin (2013)
21. Jayaram, B., Mesiar, R.: On special fuzzy implications. Fuzzy Sets and Systems 160(14), 2063–2085 (2009)
22. Kerre, E., Huang, C., Ruan, D.: Fuzzy Set Theory and Approximate Reasoning. Wu Han University Press, Wu Chang (2004)
23. Kerre, E., Nachtegael, M.: Fuzzy techniques in image processing. STUDFUZZ, vol. 52. Springer, New York (2000)
24. Klement, E., Mesiar, R., Pap, E.: Triangular norms. Kluwer Academic Publishers, Dordrecht (2000)
25. Mas, M., Monserrat, M., Torrens, J.: QL versus D-implications. Kybernetika 42, 351–366 (2006)
26. Mas, M., Monserrat, M., Torrens, J.: Two types of implications derived from uninorms. Fuzzy Sets and Systems 158, 2612–2626 (2007)
27. Mas, M., Monserrat, M., Torrens, J., Trillas, E.: A survey on fuzzy implication functions. IEEE Transactions on Fuzzy Systems 15(6), 1107–1121 (2007)
28. Massanet, S., Torrens, J.: On a new class of fuzzy implications: h-implications and generalizations. Information Sciences 181(11), 2111–2127 (2011)

29. Massanet, S., Torrens, J.: On *e*-Vertical Generated Implications. In: Melo-Pinto, P., Couto, P., Serôdio, C., Fodor, J., De Baets, B. (eds.) Eurofuse 2011. AISC, vol. 107, pp. 157–168. Springer, Heidelberg (2011)

30. Massanet, S., Torrens, J.: Intersection of Yager's implications with QL and D-implications. International Journal of Approximate Reasoning 53(4), 467–479 (2012)

31. Massanet, S., Torrens, J.: On some properties of threshold generated implications. Fuzzy Sets and Systems 205, 30–49 (2012)

32. Massanet, S., Torrens, J.: On the characterization of Yager's implications. Information Sciences 201, 1–18 (2012)

33. Massanet, S., Torrens, J.: On the vertical threshold generation method of fuzzy implication and its properties. Submitted to Fuzzy Sets and Systems (2012)

34. Massanet, S., Torrens, J.: Threshold generation method of construction of a new implication from two given ones. Fuzzy Sets and Systems 205, 50–75 (2012)

35. Ouyang, Y.: On fuzzy implications determined by aggregation operators. Information Sciences 193, 153–162 (2012)

36. Ruiz, D., Torrens, J.: Residual implications and co-implications from idempotent uninorms. Kybernetika 40, 21–38 (2004)

37. Ruiz-Aguilera, D., Torrens, J.: R-implications and S-implications from uninorms continuous in $]0,1[^2$ and their distributivity over uninorms. Fuzzy Sets and Systems 160, 832–852 (2009)

38. Shi, Y., Van Gasse, B., Kerre, E.: Fuzzy implications: Classification and a new class. In: Baczynski, M., Beliakov, G., Bustince, H., Pradera, A. (eds.) Advances in Fuzzy Implication Functions. Springer, Berlin (2013)

39. Shi, Y., Van Gasse, B., Ruan, D., Kerre, E.: On a New Class of Implications in Fuzzy Logic. In: Hüllermeier, E., Kruse, R., Hoffmann, F. (eds.) IPMU 2010, Part I. CCIS, vol. 80, pp. 525–534. Springer, Heidelberg (2010)

40. Trillas, E., Valverde, L.: On implication and indistinguishability in the setting of fuzzy logic. Management Decision Support Systems using Fuzzy Sets and Possibility Theory, 196–212 (1985)

41. Yager, R.R.: On some new classes of implication operators and their role in approximate reasoning. Information Sciences 167, 193–216 (2004)

42. Yager, R.R.: Modeling holistic fuzzy implication using co-copulas. Fuzzy Optimization and Decision Making 5, 207–226 (2006)

Fuzzy Implications: Classification and a New Class*

Yun Shi, Bart Van Gasse, and Etienne E. Kerre

Abstract. One of the most important and interesting topics in fuzzy mathematics is the study of fuzzy connectives and in particular fuzzy implications. Fuzzy implications are supposed to have at least some fundamental properties in common with the classical binary implication. Besides these fundamental properties there are many additional potential properties for fuzzy implications, among which eight are widely used in the literature. Fuzzy implications satisfying different subsets of these eight properties have been constructed and some interrelationships between these eight properties have been established. This paper aims to lay bare all the interrelationships between the eight additional properties. Where needed suitable counterexamples are provided. In our search for these counterexamples we discovered a new class of fuzzy implications that is completely determined by a fuzzy negation. For this new class we examine the conditions under which the eight properties are satisfied and we obtain the intersection with the class of strong and residual fuzzy implications.

1 Introduction

Logical operators consist an important part in the construction of classical logic. Primary logical operators in classical logic include the negation operator ¬, the conjunction operator ∧, the disjunction operator ∨ and the implication operator →. Similarly as in classical logic, logical operators play a very important role in the framework of fuzzy logic. Corresponding to the negation operator, the conjunction

Yun Shi · Bart Van Gasse · Etienne E. Kerre
Fuzziness and Uncertainty Modelling Research Unit
Department of Applied Mathematics and Computer Science,
Ghent University
e-mail: {yun.van.gent,bartvangasse}@gmail.com,
      Etienne.Kerre@ugent.be

* Dedicated to the late Prof. Dr. Da Ruan.

M. Baczyński et al. (Eds.): *Adv. in Fuzzy Implication Functions*, STUDFUZZ 300, pp. 31–51.
DOI: 10.1007/978-3-642-35677-3_2   © Springer-Verlag Berlin Heidelberg 2013

operator, the disjunction operator, and the implication operator there are negations, conjunctions, disjunctions and implications in fuzzy logic, respectively. Contrary to the operators in classical logic, the logical operators in fuzzy logic are not unique. They can be generated in different ways. These logical operators in fuzzy logic are widely studied in the literature of fuzzy mathematics. [1, 2, 3, 4, 5, 6, 7, 8, 9, 10, 12, 13, 15, 16, 17, 20, 21, 22, 23, 24, 25, 26, 29] Our study focuses mainly on the common properties of implications in fuzzy logic, and methods of generating new classes of implications.

2 Operators in Fuzzy Logic

2.1 Negations, Conjunctions and Disjunctions

The main principle in generating logical operators in fuzzy logic is that they have to coincide at the boundary points with the corresponding operators in classical logic. A *negation* N in fuzzy logic is a $[0,1] \to [0,1]$ mapping which satisfies $N(0) = 1$, $N(1) = 0$, and $x \leq y \Rightarrow N(x) \geq N(y)$, for all $x, y \in [0,1]$. Moreover, N is *strict* if it is continuous and strictly decreasing. N is *strong* if $N(N(x)) = x$, for all $x \in [0,1]$.

Strong negations are always strict. But continuous negations are not always strong. The most important strong negation in fuzzy logic is the standard negation N_0: $N_0(x) = 1 - x$, for all $x \in [0,1]$. A famous class of strong negations are the Sugeno negations N_a: there exists an $a \in]-1, +\infty[$ such that for all $x \in [0,1]$, $N_a(x) = \frac{1-x}{1+ax}$. Notice that if $a = 0$, then $N_a = N_0$. Here are two examples of classes of non-continuous negations:

$$N_A(x) = \begin{cases} 1, & \text{if } x \in A, \\ 0, & \text{if } x \notin A \end{cases}, \quad \text{for all } x \in [0,1], \tag{1}$$

where $A = [0, \alpha[$, with $\alpha \in]0,1]$, or $A = [0, \alpha]$, with $\alpha \in [0,1[$. Notice that N_A is the class of negations that take values only in $\{0,1\}$.

$$N_{A,\beta}(x) = \begin{cases} 1, & \text{if } x \in A, \\ \frac{1-x}{1+\beta x}, & \text{if } x \notin A \end{cases}, \quad \text{for all } x \in [0,1], \tag{2}$$

where $A = [0, \alpha[$, with $\alpha \in]0,1]$, or $A = [0, \alpha]$, with $\alpha \in [0,1[$, and $\beta \in]-1, +\infty[$. Notice that $N_{\{0\},\beta}$ is the class of Sugeno negations.

Widely used conjunctions in fuzzy logic are *triangular norms* (*t-norms* for short). A t-norm T is a $[0,1]^2 \to [0,1]$ mapping which satisfies $T(x,1) = x$, $y \leq z \Rightarrow T(x,y) \leq T(x,z)$, $T(x,y) = T(y,x)$ and $T(x,T(y,z)) = T(T(x,y),z)$, for all $x,y,z \in [0,1]$. Widely used disjunctions are *triangular conorms* (*t-conorms* for short). A t-conorm S is a $[0,1]^2 \to [0,1]$ mapping which satisfies $S(x,0) = x$, $y \leq z \Rightarrow S(x,y) \leq S(x,z)$, $S(x,y) = S(y,x)$ and $S(x,S(y,z)) = S(S(x,y),z)$, for all $x,y,z \in [0,1]$.

2.2 *Implications*

An *implication* I is a $[0,1]^2 \rightharpoonup [0,1]$ mapping which satisfies, for all x, x_1, x_2, y, y_1, y_2:

FI1. $x_1 < x_2 \Rightarrow I(x_1, y) \geq I(x_2, y)$ (the first place antitonicity, FA for short);
FI2. $y_1 < y_2 \Rightarrow I(x, y_1) \leq I(x, y_2)$ (the second place isotonicity, SI for short);
FI3. $I(0, x) = 1$ (dominance of falsity of antecedent, DF for short);
FI4. $I(x, 1) = 1$ (dominance of truth of consequent, DT for short);
FI5. $I(1, 0) = 0$ (boundary condition, BC for short).

Note that this definition coincides with: an implication I is a $[0,1]^2 \rightharpoonup [0,1]$ mapping satisfying FI1, FI2, $I(0,0) = 1$, $I(1,1) = 1$ and $I(1,0) = 0$. ([3], Definition 1.1.1.) Implications can be generated from negations, conjunctions and disjunctions in fuzzy logic. Let S be a t-conorm, N a negation and T a t-norm, a strong implication (S-implication for short) $I_{S,N}$ is a $[0,1]^2 \rightarrow [0,1]$ mapping which satisfies $I_{S,N}(x,y) = S(N(x), y)$, for all $x, y \in [0,1]$. A residuated implication (R-implication for short) I_T is a $[0,1]^2 \rightarrow [0,1]$ mapping which satisfies $I_T(x,y) = \sup\{t | t \in [0,1] \wedge T(x,t) \leq y\}$, for all $x, y \in [0,1]$.

Besides FI1−FI2, there are many potential properties for implications in fuzzy logic among which the following eight are the most important: for all x, y and $z \in [0,1]$,

FI6. $I(1, x) = x$ (neutrality of truth, NT for short);
FI7. $I(x, I(y, z)) = I(y, I(x, z))$ (exchange principle, EP for short);
FI8. $I(x, y) = 1 \Leftrightarrow x \leq y$ (ordering principle, OP for short);
FI9. the mapping N_I defined by $N_I(x) = I(x, 0)$, for all $x \in [0,1]$, is a strong negation (strong negation principle, SN for short);
FI10. $I(x, y) \geq y$ (consequent boundary, CB for short);
FI11. $I(x, x) = 1$ (identity, ID for short);
FI12. $I(x, y) = I(N(y), N(x))$, where N is a strong negation (contrapositive principle, CP for short);
FI13. I is a continuous mapping (continuity, CO for short).

Certain interrelationships exist between these eight properties. Next section aims to lay bare the interrelationships between these eight properties, the result of which is instrumental to propose a classification of implications.

3 Dependencies and Independencies between Properties FI6-FI13 of Implications in Fuzzy Logic

3.1 *Getting Neutrality of Truth (NT) from the Other Properties*

Theorem 1. *([3], Lemma 1.54(v), Corollary 1.57 (iii)) An implication I satisfying SN and CP w.r.t. a strong negation N satisfies NT if and only if $N_I = N$.*

In the rest of this section we consider the condition that $N_I \neq N$.

Proposition 1. *([1], Lemma 6) An implication I satisfying EP and OP always satisfies NT.*

Proposition 2. *([3], Lemma 1.56(ii)) An implication I satisfying EP and SN always satisfies NT.*

Proposition 3. *An implication I satisfying EP and CO always satisfies NT.*

Proof. Because I satisfies EP, we have for all $x \in [0,1]$,

$$I(1,N_I(x)) = I(1,I(x,0)) = I(x,I(1,0)) = I(x,0) = N_I(x). \tag{3}$$

Because I is a continuous mapping, N_I is a continuous mapping. Thus expression (3) is equivalent to $I(1,a) = a$, for all $a \in [0,1]$. Hence I satisfies NT. $\square$

Proposition 4. *There exists an implication I satisfying EP, CB, ID, CP but not NT.*

Proof. The implication I_1 stated in [6] is defined by

$$I_1(x,y) = \begin{cases} 0 & \text{if } x = 1 \text{ and } y = 0 \\ 1 & \text{else} \end{cases}, \quad \text{for all } x,y \in [0,1].$$

I_1 satisfies CB, ID and CP w.r.t. any strong negation N. However, if $x \neq 1$ then $I_1(1,x) = 1 \neq x$. Therefore I_1 does not satisfy NT. $\square$

Proposition 5. *There exists an implication I satisfying OP, SN, CB, ID, CP, CO but not NT.*

Proof. Define an implication I_2 as:

$$I_2(x,y) = \begin{cases} 1 & \text{if } x \leq y \\ \sqrt{1-(x-y)^2} & \text{if } x > y \end{cases}, \quad \text{for all } x,y \in [0,1].$$

I_2 satisfies OP, SN, CB, ID, CP w.r.t. the standard strong negation N_0, and CO. However, if $x \neq 1$ and $x \neq 0$ then $I_2(1,x) = \sqrt{2x - x^2} \neq x$. Therefore I_2 does not satisfy NT. $\square$

So for NT we have considered all possible combinations of the other seven properties that imply NT. Moreover, for all other combinations we have given a counterexample.

3.2 Getting Exchange Principle (EP) from the Other Properties

Proposition 6. *There exists an implication I satisfying NT, OP, SN, CB, ID, CP, CO but not EP.*

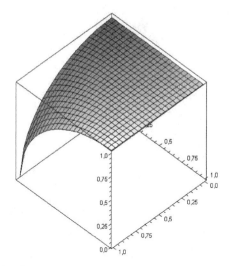

Fig. 1 The implication I_2

Proof. Define an implication I_3 as:

$$I_3(x,y) = \begin{cases} 1 & \text{if } x \leq y \\ 1-(1-y+xy)(x-y) & \text{if } x > y \end{cases}, \text{for all } x,y \in [0,1].$$

I_3 satisfies NT, OP, SN, CB, ID, CP w.r.t. the standard strong negation N_0, and CO. However, taking $x_0 = 0.5$, $y_0 = 0.9$ and $z_0 = 0.3$, we obtain $I_3(x_0, I_3(y_0, z_0)) = 0.935138$ while $I_3(y_0, I_3(x_0, z_0)) = 0.93581$. Therefore I_3 does not satisfy EP. ☐

EP is thus independent of any other of the seven properties FI6, FI8-FI13.

3.3 Getting Ordering Principle (OP) from the Other Properties

Proposition 7. *There exists an implication I satisfying NT, EP, SN, CB, ID, CP, CO but not OP.*

Proof. Given a strong negation $N(x) = \sqrt{1-x^2}$, for all $x \in [0,1]$. The S-implication I_4 generated by the t-conorm $S_L(x,y) = \min(x+y,1)$ and the strong negation N is:

$$I_4(x,y) = S_L(N(x),y) = \min(\sqrt{1-x^2}+y,1), \text{ for all } x,y \in [0,1].$$

Because I_4 is an S-implication generated from a continuous t-conorm and a strong negation, it satisfies NT, EP, SN, CB, CP w.r.t. the strong negation N and CO [8]. Moreover, for all $x \in [0,1]$, $I_4(x,x) = 1$. Therefore I_4 also satisfies ID. However, taking $x_0 = 0.6$ and $y_0 = 0.3$, we obtain $I(x_0, y_0) = 1$ while $x_0 > y_0$. Therefore I_4 does not satisfy OP. ☐

OP is thus independent of any other of the seven properties FI6, FI7, FI9-FI13.

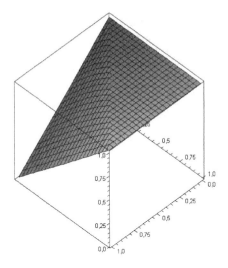

Fig. 2 The implication I_3

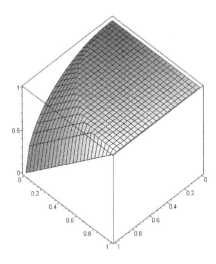

Fig. 3 The implication I_4

3.4 Getting Strong Negation Principle (SN) from the Other Properties

Proposition 8. *([3], Lemma 1.5.4(v)) An implication I satisfying NT and CP w.r.t. a strong negation N always satisfies SN. Moreover, $N_I = N$.*

Corollary 1. *An implication I satisfying EP, OP and CP w.r.t. a strong negation N always satisfies SN. Moreover, $N_I = N$.*

Proof. Straightforward from Propositions 1 and 8. □

Corollary 2. *An implication I satisfying EP, CP w.r.t. a strong negation N and CO always satisfies SN. Moreover, $N_I = N$.*

Proof. Straightforward from Propositions 3 and 8. □

Proposition 9. *([1], Lemma 14)([8], Corollary 1.1) An implication I satisfying EP, OP and CO always satisfies SN.*

Proposition 10. *([8], Table 1.1) There exists an implication satisfying NT, EP, OP, CB, ID but not SN.*

Proof. The Gödel implication

$$I_{GD}(x,y) = \begin{cases} 1, & \text{if } x \leq y \\ y, & \text{if } x > y \end{cases}, \quad \text{for all } x,y \in [0,1]. \tag{4}$$

is an R-implication generated by the continuous t-norm $T_M(x,y) = \min(x,y)$. Therefore I_{GD} satisfies NT, EP, OP, CB and ID [8]. However we have for all $x \in [0,1]$:

$$N_{I_{GD}}(x) = I_{GD}(x,0) = \begin{cases} 1, & \text{if } x = 0 \\ 0, & \text{if } x > 0 \end{cases}.$$

Thus I_{GD} does not satisfy SN. □

Proposition 11. *There exists an implication I satisfying NT, EP, CB, ID, CO but not SN.*

Proof. Given the negation $N(x) = 1 - x^2$, for all $x \in [0,1]$. The S-implication generated from the t-conorm S_L and the negation N is defined by

$$I_5(x,y) = \min(1 - x^2 + y, 1), \text{ for all } x,y \in [0,1].$$

I_5 satisfies NT, CB, ID and CO. Moreover, because I_5 is an (S,N)-implication generated from the Łukasiewicz t-conorm and the strict negation $N(x) = 1 - x^2$, it then also satisfies EP([3], Proposition 2.4.3(i)). However, we have for all $x \in [0,1]$

$$N_{I_5}(x) = I_5(x,0) = 1 - x^2$$

which is not a strong negation. Therefore I_5 does not satisfy SN. □

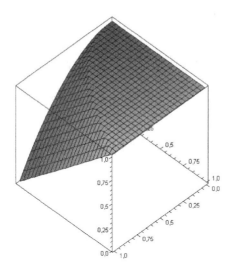

Fig. 4 The implication I_5

Proposition 12. *There exists an implication I satisfying NT, OP, CB, ID, CO but not SN.*

Proof. Define an implication I_6 as:

$$I_6(x,y) = \begin{cases} 1, & \text{if } x \leq y \\ \frac{y}{1+\sqrt{1-x}} + \sqrt{1-x}, & \text{if } x > y \end{cases}, \quad \text{for all } x,y \in [0,1].$$

I_6 satisfies NT, OP, CB, ID and CO. However, we have for all $x \in [0,1]$

$$N_{I_6}(x) = I_6(x,0) = \sqrt{1-x}$$

which is not a strong negation. Therefore I_6 does not satisfy SN □

Proposition 13. *There exists an implication I satisfying EP, CB, ID, CP but not SN.*

Proof. The implication I_1 stated in the proof of Proposition 4 satisfies EP, CB, ID and CP w.r.t. any strong negation N. However, we have

$$N_{I_1}(x) = I_1(x,0) = \begin{cases} 1, & \text{if } x < 1 \\ 0, & \text{if } x = 1 \end{cases}, \quad \text{for all } x \in [0,1],$$

which is not a strong negation. Therefore I_1 does not satisfy SN. □

Proposition 14. *There exists an implication I satisfying OP, CB, ID, CP, CO but not SN.*

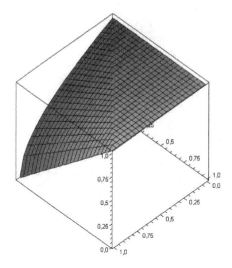

Fig. 5 The implication I_6

Proof. Define an implication I_7 as:

$$I_7(x,y) = \begin{cases} 1, & \text{if } x \le y \\ \sqrt{1-(x-y)}, & \text{if } x > y \end{cases}, \quad \text{for all } x,y \in [0,1].$$

I_7 satisfies OP, CB, ID, CP w.r.t. the standard strong negation N_0, and CO. However, we have for all $x \in [0,1]$

$$N_{I_7}(x) = I_7(x,0) = \sqrt{1-x},$$

which is not a strong negation. Therefore I_7 does not satisfy SN. $\square$

So for SN we have considered all possible combinations of the other seven properties that imply SN. Moreover, for all other combinations we have given a counterexample.

3.5 Getting Consequent Boundary (CB) from the Other Properties

Proposition 15. *([4],Lemma 1 (viii)) An implication I satisfying NT always satisfies CB.*

Corollary 3. *An implication I satisfying EP and SN always satisfies CB.*

Proof. Straightforward from Propositions 2 and 15. $\square$

Corollary 4. *An implication I satisfying EP and CO always satisfies CB.*

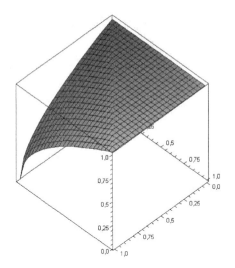

Fig. 6 The implication I_7

Proof. Straightforward from Propositions 3 and 15.   □

Proposition 16. *([1],Lemma 6) An implication I satisfying EP and OP always satisfies CB.*

Proposition 17. *There exists an implication I satisfying EP, ID, CP but not CB.*

Proof. Define an implication I_8 as:

$$I_8(x,y) = \begin{cases} 1, & \text{if } x \leq 0.5 \text{ or } y \geq 0.5 \\ 0, & \text{else} \end{cases}, \quad \text{for all } x,y \in [0,1].$$

I_8 satisfies ID and CP w.r.t. the standard strong negation N_0. However, taking $x_0 = 0.6$ and $y_0 = 0.4$, we obtain $I_8(x_0,y_0) = 0 < y_0$. Therefore I_8 does not satisfy CB.
                                                                                   □

Proposition 18. *There exists an implication I satisfying OP, SN, ID, CP, CO but not CB.*

Proof. Define an implication I_9 as:

$$I_9(x,y) = \begin{cases} 1, & \text{if } x \leq y \\ (1 - \sqrt{x-y})^2, & \text{if } x > y \end{cases}, \quad \text{for all } x,y \in [0,1].$$

I_9 satisfies OP, SN, ID, CP w.r.t. the standard strong negation N_0, and CO. However, taking $x_0 = 0.9$ and $y_0 = 0.26$, we obtain $I_9(x_0,y_0) = 0.04 < y_0$. Therefore I_9 does not satisfy CB.
                                                                                   □

So for CB we have considered all possible combinations of the other seven properties that imply CB. Moreover, for all other combinations we have given a counterexample.

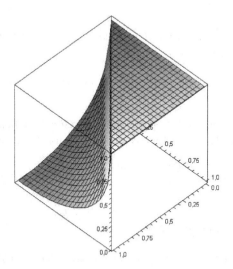

Fig. 7 The implication I_9

3.6   *Getting Identity (ID) from the Other Properties*

Proposition 19. *An implication I satisfying OP always satisfies ID.*

Proof. Straightforward.   □

Proposition 20. *There exists an implication I satisfying NT, EP, SN, CB, CP, CO but not ID.*

Proof. The Kleene-Dienes implication $I_{KD}(x,y) = \max(1-x,y)$, for all $(x,y) \in [0,1]^2$ is an S-implication generated from the t-conorm $S_{\mathbf{M}}(x,y) = \max(x,y)$ and the standard strong negation N_0. Therefore I_{KD} satisfies NT, EP, SN, CB, CP w.r.t. the standard strong negation N_0, and CO. However, for $x_0 = 0.5$, we obtain $I_{KD}(x_0,x_0) = 0.5 \neq 1$. Therefore I_{KD} does not satisfy ID.   □

So for ID we have considered all possible combinations of the other seven properties that imply ID. Moreover, for all other combinations we have given a counterexample.

3.7 Getting Contrapositive Principle (CP) from the Other Properties

Proposition 21. *([4],Lemma 1(ix)) An implication I satisfying EP and SN always satisfies CP w.r.t. the strong negation N_I.*

Proposition 22. *([1]) An implication I satisfying EP, OP and CO always satisfies CP w.r.t. the strong negation N_I.*

Proposition 23. *There exists an implication I satisfying NT, EP, OP, CB, ID but not CP.*

Proof. According to the proof of Proposition 10, the Gödel implication I_{GD} satisfies NT, EP, OP, CB and ID. However, for any strong negation N we obtain

$$I_{GD}(N(y),N(x)) = \begin{cases} 1, & \text{if } x \leq y \\ N(x), & \text{if } x > y \end{cases} \quad \text{for all } x,y \in [0,1].$$

In case that $x > y$ and $N(x) \neq y$, $I_{GD}(N(y),N(x)) \neq I_{GD}(x,y)$. Therefore I_{GD} does not satisfy CP w.r.t. any strong negation.   □

Proposition 24. *There exists an implication I satisfying NT, EP, CB, ID, CO but not CP.*

Proof. The implication I_5 stated in the proof of Proposition 11 satisfies NT, EP, CB, ID and CO. However, because for all $x \in [0,1]$, $N_{I_5}(x) = 1 - x^2$, which is strict but not a strong negation, according to Corollary 1.5.5 in [3], I_5 does not satisfy CP w.r.t. any strong negation.   □

Proposition 25. *There exists an implication I satisfying NT, OP, SN, CB, ID, CO but not CP.*

Proof. Define an implication I_{10} as:

$$I_{10}(x,y) = \begin{cases} 1, & \text{if } x \leq y \\ \frac{y+(x-y)\sqrt{1-x^2}}{x}, & \text{if } x > y \end{cases}, \quad \text{for all } x,y \in [0,1].$$

I_{10} satisfies NT, OP, SN, ID and CO. If I_{10} satisfies CP w.r.t. a strong negation N, then for all $x \in [0,1]$, we obtain $N(x) = I_{10}(1,N(x)) = I_{10}(x,0) = N_{I_{10}}(x) = \sqrt{1-x^2}$. However, take $x_0 = 0.6$ and $y_0 = 0.2$, we obtain $I_{10}(x_0,y_0) \approx 0.867$ while $I_{10}(N(y_0), N(x_0)) \approx 0.853$. Therefore I_{10} does not satisfy CP w.r.t. any strong negation N.   □

So for CP we have considered all possible combinations of the other seven properties that imply CP. Moreover, for all other combinations we have given a counterexample.

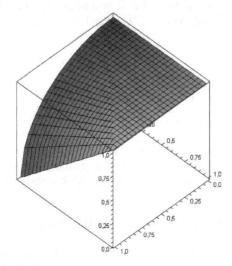

Fig. 8 The implication I_{10}

3.8 Getting Continuity (CO) from the Other Properties

Proposition 26. *There exists an implication I satisfying NT, EP, OP, SN, CB, ID, CP but not CO.*

Proof. Let N be a strong negation, the R_0-implication stated in [20] is defined as:

$$(I_{\min_0})_N(x,y) = \begin{cases} 1, & \text{if } x \leq y \\ \max(N(x),y), & \text{if } x > y \end{cases}, \quad \text{for all } x,y \in [0,1].$$

$(I_{\min_0})_N$ is the R-implication generated by the left-continuous t-norm, nilpotent minimum [9]:

$$(T_{\min_0})_N(x,y) = \begin{cases} \min(x,y), & \text{if } y > N(x) \\ 0, & \text{if } y \leq N(x) \end{cases}, \quad \text{for all } x,y \in [0,1].$$

$(I_{\min_0})_N$ satisfies NT, EP, OP, SN, CB, ID and CP w.r.t. N, and is right-continuous in the second place [20] but it is not continuous.   □

CO is thus independent of any other of the seven properties FI6-FI12.

4 A New Class of Implications in Fuzzy Logic

In the previous section we have shown the dependencies and independencies of eight important potential properties for implications in fuzzy logic. In each case where one property could not be implied from a combination of the other seven, we have given a counterexample. If we look carefully at counterexamples I_6 and I_{10} we find they actually have the same form. Let N be a negation. These two implications can be represented by I^N (with $N(x) = \sqrt{1-x}$ and $N(x) = \sqrt{1-x^2}$, respectively):

$$I^N(x,y) = \begin{cases} 1, & \text{if } x \leq y \\ \frac{(1-N(x))y}{x} + N(x), & \text{if } x > y \end{cases}, \quad \text{for all } x,y \in [0,1]. \qquad (5)$$

Remark that we obtain the Łukasiewicz implication for $N(x) = 1-x$; for $N_{\{0\}}$ we obtain the Goguen implication: $I^{N_{\{0\}}}(x,y) = I_{GG}(x,y) = \begin{cases} 1, & \text{if } x \leq y \\ y/x, & \text{if } x > y \end{cases}$, for all $x,y \in [0,1]$. If I^N is always an implication in fuzzy logic, then (5) is an interesting new class of implications, only determined by a negation. In the following we check that I^N is always an implication and have a further look at this new class.

4.1 Is I^N Defined by (5) Always an Implication?

In order to prove that I^N is always an implication in fuzzy logic we have to verify whether the mapping I^N defined by (5) takes its values in $[0,1]$ and it at least satisfies properties FI1-FI5. For convenience we first rewrite I^N as:

$$I^N(x,y) = S_\mathbf{P}(N(x), I_{GG}(x,y)), \qquad (6)$$

where $S_\mathbf{P}$ is the probabilistic sum (t-conorm): $S_\mathbf{P}(x,y) = x+y-xy$, for all $x,y \in [0,1]$. Thus straightforwardly $I^N(x,y) \in [0,1]$, and I^N satisfies F1-F5. Therefore, I^N is an implication in fuzzy logic. From (6) we can also see that I_N (seen as a function of N) is increasing. Notice that $N_{I_N} = N$, which implies that different negations result in different implications. Formula (6) can be generalised to arbitrary t-conorms and implications, but note that in that case the resulting implication does not necessarily depend on the used t-conorm, negation or implication. For example, if we start from the greatest implication, then the resulting implication will be that implication (no matter which negation and t-conorm are used). If the maximum is used as t-conorm, different negations do not always result in different implications. And also if we use the smallest negation, no new implications will be generated (no matter which t-conorm is used). Although it may very well be worthwhile to investigate this more general class, in this work we restrict ourselves to the class defined by (6).

4.2 Potential Properties of the New Class of Implications

We find out in this section whether I^N defined by (5) also satisfies the potential properties FI6-FI13. If not always, then under which conditions?

(1)NT: Straightforwardly from the definition (6), I^N always satisfies NT.
(2)EP: We obtain the following theorem:

Theorem 2. *The implication I^N defined by (5) satisfies EP if and only if N belongs to one of the following two classes of negations:*

(1)N_A defined by (1),
(2)$N_{A,\beta}$ defined by (2).

Proof. Necessity: Suppose I^N satisfies EP. We will show that if N is not of the form N_A, N must be of the form $N_{A,\beta}$. We will do this in three steps: first we will show that we can find a y_0 such that $0 < N(y_0) < y_0 < 1$. Second we prove that for $x \geq y_0$, $N(x) = \frac{1-x}{1+\beta x}$ for some fixed β. And finally we use this second step to prove that for $x < y_0$, $N(x) = 1$ or $N(x) = \frac{1-x}{1+\beta x}$.
Indeed, if I^N satisfies EP, then for all $x, y, z \in [0,1]$, $I^N(x, I^N(y, z)) = I^N(y, I^N(x, z))$. Taking $z = 0$, we obtain

$$(\forall(x,y) \in [0,1]^2)(I^N(x, N(y)) = I^N(y, N(x))). \tag{7}$$

Suppose $N \neq N_A$. Then in particular $N \neq N_{[0,1[}$. So there exists a $y_1 \in [0, 1[$ such that $N(y_1) < 1$. Now take $y_0 \in \,]\max(y_1, N(y_1)), 1[$, then $N(y_0) \leq N(y_1) < y_0 < 1$. We first show that $N(y_0) > 0$. Indeed, if $N(y_0) = 0$, then for all $x \in [0,1]$, we obtain:

$$N(x) = I^N(x, N(y_0)) = I^N(y_0, N(x)) = \begin{cases} 1, & \text{if } y_0 \leq N(x) \\ \frac{N(x)}{y_0}, & \text{if } y_0 > N(x) \end{cases},$$

$$N(x) \in \{0, 1\}, \text{ for all } x \in [0,1],$$
$$\Rightarrow N = N_A, \quad \text{for a certain } A,$$

which we have already excluded. Therefore $N(y_0) > 0$. For all $x \in [y_0, 1[, x > N(y_0)$ and $N(x) < y_0$. We obtain:

$$(7) \Rightarrow \frac{1-N(x)}{x}N(y_0) + N(x) = \frac{1-N(y_0)}{y_0}N(x) + N(y_0)$$
$$\Rightarrow \frac{1-N(x)-x}{x} = \frac{1-N(y_0)-y_0}{y_0 N(y_0)}N(x)$$

If $N(x) = 0$, then $\frac{1-N(x)-x}{x} = 0 \Rightarrow x = 1$, which we have already excluded. Therefore we obtain:

$$\frac{1-N(x)-x}{xN(x)} = \frac{1-N(y_0)-y_0}{y_0 N(y_0)}$$
$$\Rightarrow N(x) = \frac{1-x}{1+\beta x} \text{ (with } \beta = \frac{1-N(y_0)-y_0}{y_0 N(y_0)}, \beta \in \,]-1, +\infty[).$$

Now we prove, for any $x \in \,]0, y_0[$, that if $N(x) \neq 1$, then $N(x) = \frac{1-x}{1+\beta x}$. In other words that, because N is decreasing, $N = N_{A,\beta}$ defined by (2). Indeed, if $N(x) \neq 1$, then we can take y in $\,]\max(N(x), y_0), 1[$ such that $N(y) \leq x$ (this is possible because we have just proved that for $y \in [y_0, 1[, N(y) = \frac{1-y}{1+\beta y}$). We obtain:

$$(7) \Rightarrow \frac{1-N(x)-x}{xN(x)} = \frac{1-N(y)-y}{yN(y)} = \beta.$$

Thus $N(x) = \frac{1-x}{1+\beta x}$.

Sufficiency of N_A: We obtain: $I^{N_A}(x,y) = \begin{cases} 1, & \text{if } x \in A \\ I_{GG}(x,y), & \text{if } x \notin A \end{cases}$ for all $x,y \in [0,1]$.

Thus

$$I^{N_A}(x, I^{N_A}(y,z)) = \begin{cases} 1, & \text{if } x \in A \text{ or } y \in A \\ I_{GG}(x, I_{GG}(y,z)), & \text{if } x \notin A \text{ and } y \notin A \end{cases} \quad \text{for all } x,y,z \in [0,1].$$

According to [1], I_{GG} satisfies EP. Therefore I^{N_A} satisfies EP.

Sufficiency of $N_{A,\beta}$, $A = [0,\alpha[$, with $\alpha \in \,]0,1]$, or $A = [0,\alpha]$, with $\alpha \in [0,1[$: We

obtain: $I^{N_{A,\beta}}(x,y) = \begin{cases} 1, & \text{if } x \leq y \text{ or } x \in A \\ \frac{1-x+(1+\beta)y}{1+\beta x}, & \text{if } x > y \text{ and } x \notin A \end{cases}$ for all $x,y \in [0,1]$. Thus

for all $x,y,z \in [0,1]$:

$$I^{N_{A,\beta}}(x, I^{N_{A,\beta}}(y,z)) = \begin{cases} 1, & \text{if } x \in A \text{ or } y \in A \\ & \text{or } x+y+\beta xy \leq 1+z+\beta z \\ \frac{2+\beta-x-y-\beta xy+(1+\beta)^2 z}{(1+\beta x)(1+\beta y)}, & \text{else} \end{cases}$$

$$= I^{N_{A,\beta}}(y, I^{N_{A,\beta}}(x,z)). \qquad \square$$

(3) OP: We obtain the following theorem:

Theorem 3. *The implication I^N defined by (5) satisfies OP if and only if $x > 0 \Rightarrow N(x) < 1$.*

Proof. This follows from, for all $0 \leq y < x \leq 1$,

$$I^N(x,y) < 1 \Leftrightarrow \frac{1-N(x)}{x}y + N(x) < 1 \Leftrightarrow 1 - N(x) > 0 \Leftrightarrow N(x) < 1. \qquad \square$$

(4) SN: Straightforwardly $N_{I^N}(x) = I^N(x,0)$ is a strong negation if and only if N is a strong negation, because $N_{I^N} = N$.

(5) CB: Because I_{GG} satisfies CB, I^N satisfies CB according to (6).

(6) ID: We see immediately through the definition that $I^N(x,x) = 1$ for all $x \in [0,1]$.

(7) CP: We obtain the following theorem:

Theorem 4. *The implication I^N defined by (5) satisfies CP w.r.t. a strong negation N' if and only if N is a Sugeno negation N_a, $a \in \,]-1,+\infty[$, and $N' = N_a$.*

Proof. Necessity: Recall that I^N always satisfies NT. If I^N satisfies CP w.r.t. N', then according to Proposition 8, I^N also satisfies SN, and for all $x \in [0,1]$, $N'(x) = I^N(x,0) = N(x)$. Therefore, N is strong and I^N satisfies CP w.r.t. N. We obtain

$$I^N(N(y),N(x)) = I^N(x,y)$$

$$\Rightarrow (\forall x \in]0,1[)(\forall y \in]0,x[)(\frac{1-y-N(y)}{N(y)}N(x) = \frac{1-N(x)-x}{x}y)$$

$$\Rightarrow (\forall x \in]0,1[)(\forall y \in]0,x[)(\frac{1-y-N(y)}{yN(y)} = \frac{1-x-N(x)}{xN(x)})$$

$$\Rightarrow (\exists a \in [-1,+\infty])(\forall x \in]0,1[)(\frac{1-x-N(x)}{xN(x)} = a).$$

If $a = -1$ or $a = +\infty$, then $N = N_A$ defined in (1) with $A = [0,1[$ or $A = \{0\}$, which is not a strong negation. Thus $N = N_a$, which is a Sugeno negation.

Sufficiency: If $N = N_a$, then for all $x,y \in [0,1]$: $I^N(x,y) = \begin{cases} 1, & \text{if } x \le y \\ \frac{(1+a)y+1-x}{1+ax}, & \text{if } x > y \end{cases}$,

and

$$I^N(N(y),N(x)) = \begin{cases} 1, & \text{if } x \le y \\ \frac{1-y}{N(y)}N(x)+y, & \text{if } x > y \end{cases}$$

$$= \begin{cases} 1, & \text{if } x \le y \\ \frac{(1+a)y+1-x}{1+ax}, & \text{if } x > y \end{cases}.$$

Hence $I^N(x,y) = I^N(N(y),N(x))$. □

(8) CO: We obtain the following theorem:

Theorem 5. *The implication I^N defined by (5) satisfies CO if and only if N is continuous.*

Proof. It is easily verified that if N is continuous, I^N is continuous everywhere. Therefore by Corollary 1.2.2 in [3], I^N is continuous. The converse follows immediately from $I^N(x,0) = N(x)$. □

Combining the four theorems in this section and ([1], Theorem 1), we obtain the following two corollaries:

Corollary 5. *For the implication I^N defined in (5), the following four conditions are equivalent:*

i. N is a Sugeno negation N_a, $a \in]-1,+\infty[$,

ii. I^N satisfies EP and N is a continuous negation,

iii. I^N satisfies CP (w.r.t. N),

iv. I^N is conjugate with the Łukasiewicz implication I_L:

$$I_L(x,y) = \min(1-x+y,1).$$

Notice that if $a = 0$, then $N = N_0$. Then $I^N = I_L$.

Corollary 6. *An implication I^N defined by (5) satisfying EP and CO always satisfies OP.*

However, the converse of Corollary 6 is not valid. For example, the implication $I^{N_{\{0\}}} = I_{GG}$: I_{GG} satisfies OP but it is not continuous at the point $(0,0)$.

4.3 Intersection of the New Class of Implications with S- and R-implications

4.3.1 Intersection of the New Class of Implications and S-implications

Some of the implications from the new class defined by (5) are also S-implications or furthermore S-implications generated by a t-conorm and a strong negation.

Theorem 6. *The implication I^N defined by (5) is an S-implication $S(N'(x),y)$ if and only if $N = N'$ and N belongs to one of the following two classes of negations:*

(1) N_A defined by (1) with $A = [0,1[$,
(2) $N_{A,\beta}$ defined by (2).

Proof. Necessity: Because for all $x \in [0,1]$,
$N(x) = I^N(x,0) = S(N'(x),0) = N'(x)$, i.e., $N = N'$.

According to ([3], Proposition 2.4.6), any S-implication satisfies EP. Then according to Theorem 2, if I^N is an S-implication, then $N = N_A$, $A = [0,\alpha[$, with $\alpha \in]0,1]$, or $A = [0,\alpha]$, with $\alpha \in [0,1[$, or $N = N_{A,\beta}$. Nevertheless,

$$I^{N_A}(x,y) = \begin{cases} 1, & \text{if } x \leq y \text{ or } x \in A \\ \frac{y}{x}, & \text{if } x > y \text{ and } x \notin A \end{cases}, \quad \text{for all } x,y \in [0,1], \qquad (8)$$

while

$$S(N_A(x),y) = \begin{cases} 1, & \text{if } x \in A \\ y, & \text{if } x \notin A \end{cases}, \quad \text{for all } x,y \in [0,1]. \qquad (9)$$

If $A \neq [0,1[$, then we can take x and y such that $0 < y < x < 1$ and $x \notin A$. Then $S(N_A(x),y) = y \neq \frac{y}{x} = I^{N_A}(x,y)$. Thus (8)$\neq$(9) provided $A \neq [0,1[$. Therefore $I^{N_{[0,\alpha[}}$ $(\alpha < 1)$ and $I^{N_{[0,\alpha]}}$ are not S-implications.
Sufficiency of $N = N_{[0,1[}$: $I^{N_{[0,1[}}(x,y) = S(N_{[0,1[}(x),y)$ for any t-conorm S.
Sufficiency of $N = N_{A,\beta}$: Take $S(x,y) = \min(1,x+y+\beta xy)$. We can verify that S is a t-conorm (for the associativity, for all $x,y,z \in [0,1]$:

$$S(x,S(y,z)) = \min(1,x+y+z+\beta xy+\beta yz+\beta xz+\beta^2 xyz) = S(S(x,y),z),)$$

and that

$$S(N_{A,\beta}(x),y) = \begin{cases} 1, & \text{if } x \in A \text{ or } x \leq y \\ \frac{1-x+y+\beta y}{1+\beta x}, & \text{if } x \notin A \text{ and } x > y \end{cases}$$
$$= I^{N_{A,\beta}}(x,y).$$

Consequently, $I^{N_A,\beta}$ is an S-implication.   □

Connect Corollary 5 and Theorem 6 we obtain the following corollary.

Corollary 7. *For the implication I^N defined by (5), the following three conditions are equivalent:*

(1)I^N is an S-implication generated by a t-conorm and a strong negation,
(2)N is a Sugeno negation N_a, $a \in \,]-1,+\infty[$,
(3)I^N is conjugate with the Łukasiewicz implication I_L.

4.3.2 Intersection of the New Class of Implications and R-implications

Some of the implications from the new class defined by (5) are also R-implications generated by left-continuous t-norms.

Theorem 7. *The implication I^N defined by (5) is an R-implication generated by a left-continuous t-norm if and only if N belongs to one of the following two classes of negations:*

(1)a Sugeno negation N_a, $a \in \,]-1,+\infty[$,
(2)N_A defined by (1) with $A = \{0\}$.

Proof. Necessity: If I^N is an R-implication generated by a left-continuous t-norm, then according to ([8], Theorem 1.14), I^N satisfies EP and OP. According to Theorem 2, $N = N_A$ defined by (1), or $N = N_{A,\beta}$ defined by (2). According to Theorem 3, $N(x) < 1$ provided $x > 0$. Therefore $N = N_a$, or $N = N_{\{0\}}$.

Sufficiency of $N = N_a$: According to Corollary 5, if $N = N_a$, then I^N is conjugate with $I_L(x,y) = \max(x+y-1,0)$. According to ([1], Theorem 1), I^N is an R-implication.

Sufficiency of $N = N_{\{0\}}$: $I^{N_{\{0\}}} = I_{GG}$, the R-implication generated by the continuous t-norm $T_P(x,y) = xy$.   □

Notice that although $I^{N_{[0,1[}}$ is not an R-implication generated by a left-continuous t-norm, it is the R-implication I_{LR} generated by the non-left-continuous t-norm
$$T_D(x,y) = \begin{cases} \min(x,y), & \text{if } x = 1 \text{ or } y = 1 \\ 0, & \text{otherwise} \end{cases}, \quad \text{for all } x,y \in [0,1].$$

5 Conclusion Remarks

Implications play important roles in fuzzy logic. In the first part of this chapter we have analyzed the dependencies and independencies between eight potential properties FI6-FI13 of implications, found for each dependent case a proof and for each independent case a counterexample, through which we have obtained a new class of mappings defined by (5). Each of these mapping is determined by a negation only. In the second part of this chapter we have shown that the mappings defined by (5) are all implications and analyzed whether the new class of implications also satisfies the eight potential properties FI6-FI13 for implications, and if not, under which conditions the implications from this new class do satisfy these properties. Furthermore, we have also found the intersection of the new class of implications with S-implications as well as with R-implications. The new class of implications can be generalized; this broader class needs further investigation.

References

1. Baczyński, M.: Residual implications revisited. Notes on the Smets-Magrez Theorem. Fuzzy Sets and Systems 145, 267–277 (2004)
2. Baczyński, M., Drewniak, J.: Monotonic fuzzy implications. In: Szczepaniak, P.S., Lisboa, P.J.G., Kacprzyk, J. (eds.) Fuzzy Systems in Medicine, pp. 90–111. Physica, Heidelberg (2000)
3. Baczyński, M., Jayaram, B.: Fuzzy Implications. Springer, Heidelberg (2008)
4. Bustince, H., Burillo, P., Soria, F.: Automorphisms, negations and implication operators. Fuzzy Sets and Systems 134, 209–229 (2003)
5. Bustince, H., Pagola, M., Barrenechea, E.: Construction of fuzzy indices from fuzzy DI-subsethood measures: Application to the global comparison of images. Information Sciences 177, 906–929 (2007)
6. Drewniak, J.: Invariant fuzzy implications. Soft Computing 10, 506–513 (2006)
7. Fodor, J.C.: On fuzzy implication operators. Fuzzy Sets and Systems 42, 293–300 (1991)
8. Fodor, J.C., Roubens, M.: Fuzzy Preference Modelling and Multicriteria Decision Support. Kluwer Academic Publishers, Dordrecht (1994)
9. Fodor, J.C.: Contrapositive symmetry of fuzzy implications. Fuzzy Sets and Systems 69, 141–156 (1995)
10. Jayaram, B.: Rule reduction for efficient inferencing in similarity based reasoning. International Journal of Approximate Reasoning 48, 156–173 (2008)
11. Kerre, E.E.: A call for crispness in fuzzy set theory. Fuzzy Sets and Systems 29, 57–65 (1989)
12. Kerre, E.E., Nachtegael, M.: Fuzzy Techniques in Image Processing. Physica-Verlag, New York (2000)
13. Kerre, E.E., Huang, C., Ruan, D.: Fuzzy Set Theory and Approximate Reasoning. Wu Han University Press, Wu Chang (2004)
14. Klement, E.P., Mesiar, R., Pap, E.: Triangular Norms. Kluwer Academic Publishers, Netherlands (2000)
15. Klir, J., Yuan, B.: Fuzzy Sets and Fuzzy Logic, Theory and Applications. Prentice Hall, New Jersey (1995)
16. Mas, M., Monserrat, M., Torrens, J.: QL-implications versus D-implications. Kybernetika 42, 956–966 (2006)
17. Mas, M., Monserrat, M., Torrens, J., Trillas, E.: A survey on fuzzy implication functions. IEEE Transactions on Fuzzy Systems 15(6), 1107–1121 (2007)
18. Nachtegael, M., Heijmans, H., Van der Weken, D., Kerre, E.: Fuzzy Adjunctions in Mathematical Morphology. In: Proc. of JCIS 2003, North Carolina, USA, pp. 202–205 (September 2003)
19. Novák, V., Perfilieva, I., Mockor, J.: Mathematical Principles of Fuzzy Logic. Kluwer Academic Publishers, Boston (1999)
20. Pei, D.: R_0 implication: characteristics and applications. Fuzzy Sets and Systems 131, 297–302 (2002)

21. Ruan, D., Kerre, E.E.: Fuzzy implication operators and generalized fuzzy method of cases. Fuzzy Sets and Systems 54, 23–37 (1993)
22. Ruan, D., Kerre, E.E.: Fuzzy IF-THEN Rules in Computational Intelligence: Theory and Applications. Kluwer Academic Publishers, Boston (1995)
23. Ruan, D., Kerre, E.: Fuzzy IF-THEN Rules in Computational Intelligence: Theory and Applications. Kluwer Academic Publishers, Boston (2000)
24. Shi, Y., Ruan, D., Kerre, E.E.: On the characterizations of fuzzy implications satisfying $I(x, y) = I(x, I(x, y))$. Information Sciences 177(14), 2954–2970 (2007)
25. Shi, Y., Van Gasse, B., Ruan, D., Kerre, E.E.: On the first place antitonicity in QL-implications. Fuzzy Sets and Systems 159(22), 2988–3013 (2008)
26. Shi, Y., Van Gasse, B., Ruan, D., Kerre, E.E.: Interrelationships among Fuzzy Implication Axioms: Dependence versus Independence. Fuzzy Sets and Systems 161, 1388–1405 (2010)
27. Trillas, E.: Sobre funciones de negación en la teoría de conjuntos difusos. Stochastica 3(1), 47–60 (1979)
28. Trillas, E., Alsina, C., Renedo, E., Pradera, A.: On contra-symmetry and MPT conditionality in fuzzy logic. International Journal of Intelligent Systems 20, 313–326 (2005)
29. Yager, R.R.: On some new classes of implication operators and their role in approximate reasoning. Information Sciences 167, 193–216 (2004)
30. Yan, P., Chen, G.: Discovering a cover set of ARsi with hierarchy from quantitative databases. Information Sciences 173, 319–336 (2005)
31. Zadeh, L.A.: Fuzzy sets. Information and Control 8, 338–353 (1965)
32. Zadeh, L.A.: Fuzzy logic and approximate reasoning. Synthese 30, 407–428 (1975)
33. Zhang, H.Y., Zhang, W.X.: Hybrid monotonic inclusion measure and its use in measuring similarity and distance between fuzzy sets. Fuzzy Sets and Systems 160, 107–118 (2009)

A Survey of the Distributivity of Implications over Continuous T-norms and the Simultaneous Satisfaction of the Contrapositive Symmetry

Feng Qin and Michał Baczyński

Abstract. In this paper, we summarize the sufficient and necessary conditions of solutions for the distributivity equation of implication $I(x, T_1(y,z)) = T_2(I(x,y), I(x,z))$ and characterize all solutions of the system of functional equations consisting of $I(x, T_1(y,z)) = T_2(I(x,y), I(x,z))$ and $I(x,y) = I(N(y), N(x))$, when T_1 is a continuous triangular norm, T_2 is a continuous Archimedean triangular norm, I is an unknown function and N is a strong negation. We also underline that our method can be applied to other distributivity functional equations closely related to the above mentioned distributivity equation.

1 Introduction

The ability to build complex commercial and scientific fuzzy logic applications has been hampered by what is popularly known as the combinatorial rule explosion problem, which is associated with the conventional fuzzy rule configuration and its accompanying rule matrix. Since all the rules of an inference engine are exercised during every inference cycle, the number of rules directly affects the computational duration of the overall application. To reduce complexity of fuzzy "IF-THEN" rules, Combs and Andrews [10] required of the following classical tautology

$$(p \wedge q) \to r \equiv (p \to r) \vee (q \to r).$$

Feng Qin
College of Mathematics and Information Science, Nanchang Hangkong University, 330063, Nanchang, P.R. China
College of Mathematics and Information Science, Jiangxi Normal University, 330022, Nanchang, P.R. China
e-mail: qinfeng923@163.com

Michał Baczyński
Institute of Mathematics, University of Silesia, ul. Bankowa 14, 40-007 Katowice, Poland
e-mail: michal.baczynski@us.edu.pl

M. Baczyński et al. (Eds.): *Adv. in Fuzzy Implication Functions*, STUDFUZZ 300, pp. 53–72.
DOI: 10.1007/978-3-642-35677-3_3　　　© Springer-Verlag Berlin Heidelberg 2013

They refer to the left-hand side of this equivalence as an intersection rule configuration (IRC) and to its right-hand side as a union rule configuration (URC). Subsequently, there were many discussions (see [12] [18] and [11]), most of them pointed out the need for a theoretical investigation required for employing such equations, as concluded by Dick and Kandel [12], "Future work on this issue will require an examination of the properties of various combinations of fuzzy unions, intersections and implications" or by Mendel and Liang [18], "We think that what this all means is that we have to look past the mathematics of IRC⇔URC and inquire whether what we are doing when we replace IRC by URC makes sense." And then, Trillas and Alsina in [27], in the standard fuzzy theory, turned the about requirement into the functional equation $I(T(x,y),z) = S(I(x,z),I(y,z))$ and obtained all solutions of T when I are special cases of R-implications, S-implications and QL-implications, respectively. Along the lines, Balasubramaniam and Rao in [8] investigated the other three functional equations interrelated with this equation. In order to study it in more general case, Ruiz-Aguilera [24], [25] and Qin [23], in their own papers, generalized the above equations into uninorms.

On the other hand, from fuzzy logical angle, Türksen et al. [29] posed and discussed the equation

$$I(x, T(y,z)) = T(I(x,y), I(x,z)), \qquad x,y,z \in [0,1], \tag{1}$$

and then, got the necessary conditions for a fuzzy implication I to satisfy Eq. (1) when $T = T_P$. Later, Baczyński [1] generalized some Türksen's results into strict t-norm T and obtained the sufficient and necessary conditions of functional equations consisting of Eq. (1) and the following equation $I(x, I(y,z)) = I(T(x,y),z)$. Moreover, he also studied in [2] the system of functional equations composed of Eq. (1) and the following equation

$$I(x,y) = I(N(y), N(x)), \qquad x,y \in [0,1]. \tag{2}$$

After this, Yang and Qin in [30] got the full characterizations of the system of functional equations composed of Eq. (1) and Eq. (2) when T is a strict t-norm. Recently, many people (see [6], [3] [22]) investigate again the distributivity of fuzzy implications over nilpotent or strict t-norms or t-conorms. Specially, we in [21], in the most general case, explored and got the sufficient and necessary conditions of solutions for the distributivity equation of implication

$$I(x, T_1(y,z)) = T_2(I(x,y), I(x,z)), \quad x,y,z \in [0,1], \tag{3}$$

And then, we in [19], characterized solutions of the system of functional equations consisting of Eq. (2) and Eq. (3). Along the above line, in this paper, we summarize the sufficient and necessary conditions of solutions for Eq. (3) and the system of functional equations consisting of Eq. (2) and Eq. (3), when T_1 is a continuous triangular norm, T_2 is a continuous and Archimedean triangular norm, I is an unknown function and N is a strong negation. We also underline that our method can be applied to the three other distributivity functional equations closely related to the above-mentioned functional equation.

The paper is organized as follows. In Section 2 we present some results concerning basic fuzzy logic connectives employed in the sequel. In Section 3 we investigate Eq. (3) when T_1 is a continuous t-norm, T_2 is a continuous Archimedean t-norm and I is an unknown function. In Section 4 we will describe all solutions of Eq. (3) when T_2 is a strict t-norm. In Section 5 we will repeat these investigations but with the assumption that T_2 is a nilpotent t-norm.

In section 6 we discuss the system of functional equations consisting of Eq. (2) and Eq. (3) when both t-norms T_1, T_2 are continuous and Archimedean. In section 7 we briefly discuss the solutions when $T_1 = T_M$ minimum t-norm. In section 8 we investigate the system of functional equations consisting of Eq. (2) and Eq. (3) when T_1 is a continuous t-norm and T_2 is a strict t-norm. In section 9 we repeat the above investigations but with the assumption that T_2 is a nilpotent t-norm.

2   Preliminaries

Definition 2.1 ([13], [14], [15]). A binary function $T \colon [0,1]^2 \to [0,1]$ is called a *triangular norm* (t-norm for short), if it fulfills, for every $x, y, z \in [0,1]$, the following conditions:

1. $T(x,y) = T(y,x)$,   (commutativity)
2. $T(T(x,y),z) = T(x,T(y,z))$,   (associativity)
3. $T(x,y) \leq T(x,z)$, whenever $y \leq z$,   (monotonicity)
4. $T(x,1) = x$.   (boundary condition)

Please note that an element $a \in [0,1]$ is called and *idempotent element* of T if $T(a,a) = a$.

Definition 2.2 ([14], [15]). A t-norm T is said to be

1. *Archimedean*, if for every $x, y \in (0,1)$, there exists some $n \in \mathbf{N}$ such that $x_T^n < y$, where $x_T^n = T(\underbrace{x, x, \cdots, x}_{n \text{ times}})$;

2. *strict*, if it is continuous and strictly monotone, i.e., $T(x,y) < T(x,z)$ whenever $x \in (0,1]$ and $y < z$;
3. *nilpotent*, if it is continuous and if for each $x \in (0,1)$ there exists some $n \in \mathbf{N}$ such that $x_T^n = 0$.

Remark 2.3. 1. A continuous t-norm T is Archimedean if and only if it holds $T(x,x) < x$ for all $x \in (0,1)$ (see Proposition 5.1.2 in [14]).
2. If T is strict or nilpotent, then it must be Archimedean. The converse is also true when it is continuous (see Theorem 2.18 in [15]).

Theorem 2.4 ([15], [17]). *For a function $T \colon [0,1]^2 \to [0,1]$, the following statements are equivalent:*

(i) T is a continuous Archimedean t-norm.

(ii) T has a continuous additive generator, i.e., there exists a continuous, strictly decreasing function $t\colon [0,1] \to [0,\infty]$ with $t(1) = 0$, which is uniquely determined up to a positive multiplicative constant, such that

$$T(x,y) = t^{-1}(\min(t(x) + t(y), t(0))), \qquad x,y \in [0,1]. \tag{4}$$

Remark 2.5 ([15]).

1. A t-norm T is strict if and only if each continuous additive generator t of T satisfies $t(0) = \infty$.
2. A t-norm T is nilpotent if and only if each continuous additive generator t of T satisfies $t(0) < \infty$.

Theorem 2.6 ([14], [15]). *T is a continuous t-norm, if and only if*

1. *$T = T_M$ (the minimum t-norm), or*
2. *T is continuous Archimedean, or*
3. *there exists a family $\{[a_m, b_m], T_m\}_{m \in A}$ such that T is the ordinal sum of this family denoted by $T = (\langle a_m, b_m, T_m \rangle)_{m \in A}$. In other words,*

$$T(x,y) = \begin{cases} a_m + (b_m - a_m) T_m(\frac{x - a_m}{b_m - a_m}, \frac{y - a_m}{b_m - a_m}), & \text{if } x,y \in [a_m, b_m], \\ \min(x,y), & \text{otherwise,} \end{cases} \tag{5}$$

where $\{[a_m, b_m]\}_{m \in A}$ is a family of non-over lapping, closed, proper subintervals of $[0,1]$ with each T_m being a continuous Archimedean t-norm, and A is a finite or countable infinite index set. For every $m \in A$, $[a_m, b_m]$ is called generating subinterval of T, and T_m the corresponding generating t-norm on $[a_m, b_m]$ of T.

In the literature we can find various definitions of fuzzy implications (see [6], [9], [15], [28]). But, in this article, we will use the following one, which is equivalent to the definition introduced by Fodor and Roubens (see [13]).

Definition 2.7 ([6], [5], [13]). A function $I\colon [0,1]^2 \to [0,1]$ is called a *fuzzy implication*, if I fulfills the following conditions:

I1: I is decreasing with respect to the first variable;
I2: I is increasing with respect to the second one;
I3: $I(0,0) = I(0,1) = I(1,1) = 1, I(1,0) = 0$.

In virtue of the above definition, it is obvious that each fuzzy implication satisfy $I(0,x) = I(x,1) = 1$ for all $x \in [0,1]$. But we can say nothing about the value of $I(x,0)$ and $I(1,x)$ for all $x \in (0,1)$.

Definition 2.8. A continuous function $N\colon [0,1] \to [0,1]$ is called a *strong negation*, if it is strictly decreasing, involutive and satisfies $N(0) = 1$ and $N(1) = 0$. Specially, when $N(x) = 1 - x$, we call it the *standard negation*, denoted by N_0.

Theorem 2.9 (Trillas [26]). *For a function $N\colon [0,1] \to [0,1]$ the following statements are equivalent:*

(i) N is a strong negation.
(ii) There exists an increasing bijection $\varphi\colon [0,1] \to [0,1]$ such that

$$N(x) = \varphi^{-1}(1 - \varphi(x)), \qquad x \in [0,1]. \tag{6}$$

Finally, let us mention that in the proofs of main results the solutions of the additive Cauchy functional equation (see [16]) and of similar functional equations are very important. For these facts with the proofs see [7] and [3].

3 First Solutions to Eq. (3) When T_1 Is a Continuous T-norm

In this section, we present first characterizations of function I satisfying Eq. (3) when T_1 is a continuous t-norm and T_2 is a continuous Archimedean t-norm. For the proofs see [21, Section 3].

Lemma 3.1. *Let T_1 be a continuous t-norm, T_2 be a continuous Archimedean t-norm and $I\colon [0,1]^2 \to [0,1]$ be a binary function. If the triple of functions (T_1,T_2,I) satisfies Eq. (3) for all $x,y,z \in [0,1]$, then either $I(x,y) = 0$ or $I(x,y) = 1$ hold for every $x \in [0,1]$ and every $y \notin \cup_{m\in A}(a_m,b_m)$, i.e., y is an idempotent element of T_1, where $m \in A$ and (a_m,b_m) are the symbols in Theorem 2.6.*

Lemma 3.2. *Let T_1 be a continuous t-norm, T_2 be a continuous Archimedean t-norm, $I\colon [0,1]^2 \to [0,1]$ be a binary function and y_0 be a fixed idempotent element of T_1. If the triple of functions (T_1,T_2,I) satisfies Eq. (3) for all $x,y,z \in [0,1]$, then*

(i) if it holds $I(x,y_0) = 0$ for some $x \in [0,1]$, then it follows $I(x,z) = 0$ for any $z \leq y_0$.
(ii) if it holds $I(x,y_0) = 1$ for some $x \in [0,1]$, then it follows $I(x,z) = 1$ for any $z \geq y_0$.

According to the above analysis, we have the following result.

Lemma 3.3. *Let T_1 be a continuous t-norm, T_2 be a continuous Archimedean t-norm, $I\colon [0,1]^2 \to [0,1]$ be a binary function, y_1,y_2 be two different idempotent elements of T_1 and assume that the triple of functions (T_1,T_2,I) satisfies Eq. (3) for all $x,y,z \in [0,1]$. If both $I(x,y_1) = 0$ and $I(x,y_2) = 1$ are simultaneously true for some $x \in [0,1]$, then we must have $y_1 < y_2$.*

For any given continuous t-norm T_1, binary function I, and fixed $x \in [0,1]$, we define

$$U_{(T_1,I,x)} = \{y \in [0,1] \mid I(x,y) = 0 \text{ and } y \text{ is an idempotent element of } T_1\},$$
$$\mu_{(T_1,I,x)} = \sup U_{(T_1,I,x)},$$

and

$$V_{(T_1,I,x)} = \{y \in [0,1] \mid I(x,y) = 1 \text{ and } y \text{ is an idempotent element of } T_1\},$$
$$\nu_{(T_1,I,x)} = \inf V_{(T_1,I,x)}.$$

We stipulate here that $\sup \emptyset = 0$ and $\inf \emptyset = 1$. Obviously, under conditions of Lemma 3.1, it holds from Lemma 3.3 that $\mu_{(T_1,I,x)} \leq \nu_{(T_1,I,x)}$ for any T_1, I and

$x \in [0,1]$. Since T_1 is continuous and monotone, it follows that both $\mu_{(T_1,I,x)}$ and $\nu_{(T_1,I,x)}$ are idempotent elements of T_1, too. Now, by the order between $\mu_{(T_1,I,x)}$ and $\nu_{(T_1,I,x)}$, we need to consider two cases: $\mu_{(T_1,I,x)} = \nu_{(T_1,I,x)}$ and $\mu_{(T_1,I,x)} < \nu_{((T_1,I,x)}$.

Theorem 3.4. *Let T_1 be a continuous t-norm, T_2 be a continuous Archimedean t-norm, $I : [0,1]^2 \to [0,1]$ be a binary function and assume that $\mu_{(T_1,I,x)} = \nu_{(T_1,I,x)}$ for some fixed $x \in [0,1]$. Then the following statements are equivalent:*

(i) The triple of functions $(T_1, T_2, I(x,\cdot))$ satisfies Eq. (3) for any $y,z \in [0,1]$;
(ii) The vertical section $I(x,\cdot)$ has the following forms:

   a. If $\mu_{(T_1,I,x)} \in U_{(T_1,I,x)}$, then

$$I(x,y) = \begin{cases} 0, & \text{if } y \le \mu_{(T_1,I,x)}, \\ 1, & \text{if } y > \mu_{(T_1,I,x)}, \end{cases} \qquad y \in [0,1]. \tag{7}$$

   b. If $\nu_{(T_1,I,x)} \in V_{(T_1,I,x)}$, then

$$I(x,y) = \begin{cases} 0, & \text{if } y < \nu_{(T_1,I,x)}, \\ 1, & \text{if } y \ge \nu_{(T_1,I,x)}, \end{cases} \qquad y \in [0,1]. \tag{8}$$

Remark 3.5. Note that, in Theorem 3.4, if $\mu_{(T_1,I,x)} = 0$, then it means $U_{(T_1,I,x)} = \{0\}$ or $U_{(T_1,I,x)}$ is empty. If $U_{(T_1,I,x)} = \{0\}$, then $I(x,y)$ has the form in Eq. (7); if $U_{(T_1,I,x)}$ is empty, then $I(x,0) = 1$ and $\nu_{(T_1,I,x)} = 0$, which implies that $I(x,y)$ has the form in Eq. (8), i.e., $I(x,\cdot) = 1$. The case $\mu_{(T_1,I,x)} = 1$ is similar and includes the solution $I(x,\cdot) = 0$.

Corollary 3.6. *Let $T_1 = T_M$ (the minimum t-norm), T_2 be a continuous Archimedean t-norm and $I : [0,1]^2 \to [0,1]$ be a binary function. Then the following statements are equivalent:*

(i) The triple of functions (T_M, T_2, I) satisfies Eq. (3) for all $x,y,z \in [0,1]$.
(ii) For any $x \in [0,1]$ there exists a constant $c_x \in [0,1]$ such that the vertical section $I(x,\cdot)$ has one of the following forms

$$I(x,y) = \begin{cases} 0, & \text{if } y \le c_x, \\ 1, & \text{if } y > c_x, \end{cases} \qquad y \in [0,1], \tag{9}$$

$$I(x,y) = \begin{cases} 0, & \text{if } y < c_x, \\ 1, & \text{if } y \ge c_x, \end{cases} \qquad y \in [0,1]. \tag{10}$$

Now, let us consider the case $\mu_{(T_1,I,x)} < \nu_{(T_1,I,x)}$.

Lemma 3.7. *Let T_1 be a continuous t-norm, T_2 be a continuous Archimedean t-norm, $I : [0,1]^2 \to [0,1]$ be a binary function, and fix arbitrarily $x \in [0,1]$. If $\mu_{(T_1,I,x)} < \nu_{(T_1,I,x)}$ and the triple of functions $(T_1, T_2, I(x,\cdot))$ satisfies Eq. (3), for all $y,z \in [0,1]$, then there exists some $\alpha_0 \in A$ such that $[\mu_{(T_1,I,x)}, \nu_{(T_1,I,x)}] = [a_{\alpha_0}, b_{\alpha_0}]$,* *where A and $[a_{\alpha_0}, b_{\alpha_0}]$ are the index set and the subinterval in Theorem 2.6, respectively.*

Remark 3.8. So far, we have obtained the fact that

$$I(x,y) = \begin{cases} 0, & \text{if } y < \mu_{(T_1,I,x)}, \\ 1, & \text{if } y > \nu_{(T_1,I,x)}, \end{cases}$$

for any fixed $x \in [0,1]$, when $\mu_{(T_1,I,x)} < \nu_{(T_1,I,x)}$. But we do say nothing about the value of $I(x,y)$ for any $y \in [\mu_{(T_1,I,x)}, \nu_{(T_1,I,x)}]$. We will solve this problem in the next two sections, considering the different assumptions on t-norm T_2.

4 Solutions to Eq. (3) When T_1 Is a Continuous T-norm and T_2 Is a Strict T-norm

In this section, we characterize the fuzzy implication I satisfying Eq. (3) when T_1 is a continuous t-norm and T_2 is a strict t-norm. From Lemmas 3.1, 3.2, 3.3, 3.7 and Theorem 3.4, it is enough to consider the case $\mu_{(T_1,I,x)} < \nu_{(T_1,I,x)}$ and T_2 is a strict t-norm. For the proofs see [21, Section 4].

Theorem 4.1. *Let T_1 be a continuous t-norm, T_2 be a strict t-norm, $I: [0,1]^2 \to [0,1]$ be a binary function and fix arbitrarily $x \in [0,1]$. If $\mu_{(T_1,I,x)} < \nu_{(T_1,I,x)}$ and the corresponding generating t-norm T_{α_0} of T_1 on the generating subinterval $[\mu_{(T_1,I,x)}, \nu_{(T_1,I,x)}] = [a_{\alpha_0}, b_{\alpha_0}]$ is strict, then the following statements are equivalent:*

(i) The triple of functions $(T_1, T_2, I(x,\cdot))$ satisfies Eq. (3) for all $y, z \in [0,1]$.

(ii) T_1 admits the representation (5), and there exist continuous, strictly decreasing functions $t_{\alpha_0}, t_2: [0,1] \to [0,\infty]$ with $t_{\alpha_0}(1) = t_2(1) = 0$, $t_{\alpha_0}(0) = t_2(0) = \infty$, which are uniquely determined up to positive multiplicative constants, such that the corresponding generating t-norm T_{α_0} of T_1 on the generating subinterval $[a_{\alpha_0}, b_{\alpha_0}]$ and T_2 admit the representation (4) with t_{α_0}, t_2, respectively, and for the mentioned above $x \in [0,1]$, the vertical section $I(x,\cdot)$ has one of the following forms:

$$I(x,y) = \begin{cases} 0, & \text{if } y \in [0, a_{\alpha_0}], \\ 1, & \text{if } y \in (a_{\alpha_0}, 1], \end{cases} \tag{11}$$

$$I(x,y) = \begin{cases} 0, & \text{if } y \in [0, b_{\alpha_0}), \\ 1, & \text{if } y \in [b_{\alpha_0}, 1], \end{cases} \tag{12}$$

$$I(x,y) = \begin{cases} 0, & \text{if } y \in [0, a_{\alpha_0}], \\ t_2^{-1}(c_x t_{\alpha_0}(\frac{y - a_{\alpha_0}}{b_{\alpha_0} - a_{\alpha_0}})), & \text{if } y \in [a_{\alpha_0}, b_{\alpha_0}], \\ 1, & \text{if } y \in [b_{\alpha_0}, 1], \end{cases} \tag{13}$$

with a certain $c_x \in (0,\infty)$, uniquely determined up to a positive multiplicative constant depending on constants for t_{α_0} and t_2, i.e., if $t'_{\alpha_0}(y) = a t_{\alpha_0}(y)$,

$t'_2(y) = bt_2(y)$ for all $y \in [0,1]$ and some $a,b \in (0,\infty)$, and we assume that $t_2^{-1}(c_x t_{\alpha_0}(\frac{y-a_{\alpha_0}}{b_{\alpha_0}-a_{\alpha_0}})) = t_2'^{-1}(c'_x t'_{\alpha_0}(\frac{y-a_{\alpha_0}}{b_{\alpha_0}-a_{\alpha_0}}))$, $y \in [a_{\alpha_0}, b_{\alpha_0}]$, then $c'_x = \frac{b}{a}c_x$.

Remark 4.2. In particular, when T_1 and T_2 are strict t-norms, then there are only two idempotent elements of T_1: 0 and 1. Therefore, for any fixed $x \in [0,1]$ we get three possible cases:

- If $\mu_{(T_1,I,x)} = v_{(T_1,I,x)} = 0$, then the only solution is $I(x,y) = 0$ for all $y \in [0,1]$ (it follows from Theorem 3.4).
- If $\mu_{(T_1,I,x)} = v_{(T_1,I,x)} = 1$, then the only solution is $I(x,y) = 1$ for all $y \in [0,1]$ (it follows again from Theorem 3.4).
- If $\mu_{(T_1,I,x)} < v_{(T_1,I,x)}$, then $\mu_{(T_1,I,x)} = 0$ and $v_{(T_1,I,x)} = 1$ and from Theorem 4.1 we get three possible solutions:

$$I(x,y) = \begin{cases} 0, & \text{if } y = 0, \\ 1, & \text{if } y > 0, \end{cases} \tag{14}$$

$$I(x,y) = \begin{cases} 0, & \text{if } y < 1, \\ 1, & \text{if } y = 1, \end{cases} \tag{15}$$

$$I(x,y) = t_2^{-1}(c_x t_1(y)), \tag{16}$$

where $t_1, t_2 : [0,1] \to [0,\infty]$ are continuous, strictly decreasing functions with $t_1(1) = t_2(1) = 0, t_1(0) = t_2(0) = \infty$.

Please note that these solutions are exactly the same as in the results already presented by Baczyński in [1], [2].

Theorem 4.3. *Let T_1 be a continuous t-norm, T_2 be a strict t-norm, $I: [0,1]^2 \to [0,1]$ be a binary function and fix arbitrarily $x \in [0,1]$. If $\mu_{(T_1,I,x)} < v_{(T_1,I,x)}$ and the corresponding generating t-norm T_{α_0} of T_1 on the generating subinterval $[\mu_{(T_1,I,x)}, v_{(T_1,I,x)}] = [a_{\alpha_0}, b_{\alpha_0}]$ is nilpotent, then the following statements are equivalent:*

(i) The triple of functions $(T_1, T_2, I(x,\cdot))$ satisfies Eq. (3) for all $y,z \in [0,1]$.

(ii) T_1 admits the representation (5), and there exist continuous, strictly decreasing functions $t_{\alpha_0}, t_2 : [0,1] \to [0,\infty]$ with $t_{\alpha_0}(1) = t_2(1) = 0, t_{\alpha_0}(0) < \infty, t_2(0) = \infty$, which are uniquely determined up to positive multiplicative constants, such that the corresponding generating t-norm T_{α_0} of T_1 on the generating subinterval $[a_{\alpha_0}, b_{\alpha_0}]$ and T_2 admit the representation (4) with t_{α_0}, t_2, respectively, and for the mentioned above $x \in [0,1]$, the vertical section has the following form

$$I(x,y) = \begin{cases} 0, & \text{if } y \in [0, b_{\alpha_0}), \\ 1, & \text{if } y \in [b_{\alpha_0}, 1]. \end{cases} \tag{17}$$

Remark 4.4. In particular, when T_1 is a nilpotent t-norm, then there are only two idempotent elements: 0 and 1. Therefore, since T_2 is a strict t-norm, then for any fixed $x \in [0,1]$ we get three possible cases:

- If $\mu_{(T_1,I,x)} = \nu_{(T_1,I,x)} = 0$, then the only solution is $I(x,y) = 0$ for all $y \in [0,1]$ (it follows from Theorem 3.4).
- If $\mu_{(T_1,I,x)} = \nu_{(T_1,I,x)} = 1$, then the only solution is $I(x,y) = 1$ for all $y \in [0,1]$ (it follows again from Theorem 3.4).
- If $\mu_{(T_1,I,x)} < \nu_{(T_1,I,x)}$, then $\mu_{(T_1,I,x)} = 0$ and $\nu_{(T_1,I,x)} = 1$ and from Theorem 4.3 we get the only solution:

$$I(x,y) = \begin{cases} 0, & \text{if } y < 1, \\ 1, & \text{if } y = 1. \end{cases} \tag{18}$$

Please note that these solutions are exactly the same as in the results already presented by Qin and Baczyński in [20, Theorem 4.2].

Next, let us consider continuous solutions for Eq. (3).

Theorem 4.5. *Let T_1 be a continuous t-norm, T_2 be a strict t-norm and $I: [0,1]^2 \to [0,1]$ be a continuous binary function. Then the following statements are equivalent:*

(i) The triple of functions (T_1, T_2, I) satisfies Eq. (3) for all $x,y,z \in [0,1]$.
(ii) T_1 admits the representation (5), and there exist two constants $a < b \in [0,1]$ such that $\mu_{(T_1,I,x)} = a$, $\nu_{(T_1,I,x)} = b$ for all $x \in [0,1]$, and there exist two continuous, strictly decreasing functions $t_a, t_2 \colon [0,1] \to [0,\infty]$ with $t_a(1) = t_2(1) = 0$, $t_a(0) = t_2(0) = \infty$, which are uniquely determined up to a positive multiplicative constant, such that the corresponding generating t-norm T_a of T_1 on the generating subinterval $[a,b]$ and T_2 admit the representation (4) with t_a and t_2 respectively, and $I = 0$ or, $I = 1$ or, there exists a continuous function $c \colon [0,1] \to (0,\infty)$, uniquely determined up to a positive multiplicative constant depending on constants for t_a and t_2, such that I has the form

$$I(x,y) = \begin{cases} 0, & \text{if } y \in [0,a], \\ t_2^{-1}(c(x)t_a(\frac{y-a}{b-a})), & \text{if } y \in [a,b], \\ 1, & \text{if } y \in [b,1], \end{cases} \tag{19}$$

for all $x,y \in [0,1]$.

Corollary 4.6. *If T_1 is a continuous t-norm and T_2 is a strict t-norm, then there are no continuous solutions I of Eq. (3) which satisfy axiom I3.*

Proof. Let I be a continuous binary function satisfying axiom I3. By Theorem 4.5, I must have the form in Eq. (19). But in this case we get $I(0,0) = 0$ because of $a \geq 0$, which is a contradiction. □

Remark 4.7. From the above proof it is obvious that we need to look for the solutions of Eq. (3), which are continuous except at the point $(0,0)$. However, substituting $x = 0$ and $z = 0$ in Eq. (3), we know that it follows that $I(0, T_1(y,0)) = T_2(I(0,y), I(0,0))$ for all $y \in [0,1]$. That is, $I(0,0) = T_2(I(0,y), 1)$, which implies $I(0,0) = I(0,y)$. So it holds $I(0,y) = 1$ for all $y \in [0,1]$. Thus we have proven the fact that it holds $I(0,y) = 1$ for all $y \in [0,1]$ if the binary function I satisfying axiom I3 is a solution of Eq. (3). On the other hand, applying the method of Theorem 4.3 in [22], we can further know that there exist only solutions which are continuous except at the vertical section $I(0,y) = 1$ for $y \in [0,a]$.

Theorem 4.8. *Let T_1 be a continuous t-norm, T_2 be a strict t-norm, $I: [0,1]^2 \to [0,1]$ be a continuous fuzzy implication except at the vertical section $I(0,y) = 1$ for $y \in [0,a]$. Then the following statements are equivalent:*

(i) The triple of functions (T_1, T_2, I) satisfies Eq. (3) for all $x, y, z \in [0,1]$.

(ii) T_1 admits the representation (5), there exist two constants $a < b \in [0,1]$ such that $\mu_{(T_1,I,x)} = a$, $\nu_{(T_1,I,x)} = b$ for all $x \in [0,1]$, and there exist continuous, strictly decreasing functions $t_a, t_2: [0,1] \to [0,\infty]$ with $t_a(1) = t_2(1) = 0$, $t_a(0) = t_2(0) = \infty$, which are uniquely determined up to positive multiplicative constants, such that the corresponding generating t-norm T_a of T_1 on the generating subinterval $[a,b]$ and T_2 admit the representation (4) with t_a and t_2 respectively, and there exists a continuous and increasing function $c: (0,1] \to (0,\infty), c(0) = 0$, uniquely determined up to a positive multiplicative constant depending on constants for t_a and t_2, such that I has the form

$$I(x,y) = \begin{cases} 1, & \text{if } x = 0, y \in [0,a], \\ 0, & \text{if } x \neq 0, y \in [0,a], \\ t_2^{-1}(c(x)t_a(\frac{y-a}{b-a})), & \text{if } y \in [a,b], \\ 1, & \text{if } y \in [b,1]. \end{cases} \tag{20}$$

5 Solutions to Eq. (3) When T_1 Is a Continuous T-norm and T_2 Is a Nilpotent T-norm

In this section, we characterize the fuzzy implication I satisfying Eq. (3) when T_1 is a continuous t-norms and T_2 is a nilpotent t-norm. From Lemmas 3.1, 3.2, 3.3, 3.7 and Theorem 3.4, it is enough to consider the case $\mu_{(T_1,I,x)} < \nu_{(T_1,I,x)}$ and T_2 is a nilpotent t-norm. For the proofs see [21, Section 5].

Theorem 5.1. *Let T_1 be a continuous t-norm, T_2 be a nilpotent t-norm, $I: [0,1]^2 \to [0,1]$ be a binary function and fix $x \in [0,1]$. If $\mu_{(T_1,I,x)} < \nu_{(T_1,I,x)}$ and the corresponding generating t-norm T_{α_0} of T_1 on the generating subinterval $[\mu_{(T_1,I,x)}, \nu_{(T_1,I,x)}] = [a_{\alpha_0}, b_{\alpha_0}]$ is strict, then the following statements are equivalent:*

(i) The triple of functions $(T_1, T_2, I(x, \cdot))$ satisfies Eq. (3) for all $y, z \in [0,1]$.

(ii) T_1 admits the representation (5), and there exist continuous, strictly decreasing functions $t_{\alpha_0}, t_2: [0,1] \to [0,\infty]$ with $t_{\alpha_0}(1) = t_2(1) = 0$, $t_{\alpha_0}(0) = \infty, t_2(0) < \infty$, which are uniquely determined up to positive multiplicative constants, such that the corresponding generating t-norm T_{α_0} of T_1 on the generating subinterval $[a_{\alpha_0}, b_{\alpha_0}]$ and T_2 admit the representation (4) with t_{α_0}, t_2, respectively, and for the mentioned above $x \in [0,1]$, the vertical section $I(x, \cdot)$ has one of the following forms:

$$I(x,y) = \begin{cases} 0, & \text{if } y \in [0, a_{\alpha_0}], \\ 1, & \text{if } y \in (a_{\alpha_0}, 1]. \end{cases} \tag{21}$$

$$I(x,y) = \begin{cases} 0, & \text{if } y \in [0, b_{\alpha_0}), \\ 1, & \text{if } y \in [b_{\alpha_0}, 1], \end{cases} \tag{22}$$

$$I(x,y) = \begin{cases} 0, & \text{if } y \in [0, a_{\alpha_0}], \\ t_2^{-1}(\min(c_x t_{\alpha_0}(\frac{y - a_{\alpha_0}}{b_{\alpha_0} - a_{\alpha_0}}), t_2(0))), & \text{if } y \in [a_{\alpha_0}, b_{\alpha_0}], \\ 1, & \text{if } y \in [b_{\alpha_0}, 1], \end{cases} \tag{23}$$

with a certain $c_x \in (0, \infty)$, uniquely determined up to a positive multiplicative constant depending on constants for t_{α_0} and t_2, i.e., if $t'_{\alpha_0}(y) = a t_{\alpha_0}(y)$, $t'_2(y) = b t_2(y)$ for all $y \in [0,1]$ and some $a, b \in (0, \infty)$, and we assume that $t_2^{-1}(\min(c_x t_{\alpha_0}(\frac{y - a_{\alpha_0}}{b_{\alpha_0} - a_{\alpha_0}}), t_2(0))) = t'^{-1}_2(\min(c'_x t'_{\alpha_0}(\frac{y - a_{\alpha_0}}{b_{\alpha_0} - a_{\alpha_0}}), t'_2(0))), y \in [a_{\alpha_0}, b_{\alpha_0}]$, then $c'_x = \frac{b}{a} c_x$.

Remark 5.2. In particular, when T_1 is a strict t-norm, then there are only two idempotent elements: 0 and 1. Therefore, if T_2 is a nilpotent t-norm, then for any fixed $x \in [0,1]$ we get three possible cases:

- If $\mu_{(T_1, I, x)} = \nu_{(T_1, I, x)} = 0$, then the only solution is $I(x,y) = 0$ for all $y \in [0,1]$ (it follows from Theorem 3.4).
- If $\mu_{(T_1, I, x)} = \nu_{(T_1, I, x)} = 1$, then the only solution is $I(x,y) = 1$ for all $y \in [0,1]$ (it follows again from Theorem 3.4).
- If $\mu_{(T_1, I, x)} < \nu_{(T_1, I, x)}$, then $\mu_{(T_1, I, x)} = 0$ and $\nu_{(T_1, I, x)} = 1$ and from Theorem 5.1 we get three possible solutions:

$$I(x,y) = \begin{cases} 0, & \text{if } y = 0, \\ 1, & \text{if } y > 0, \end{cases} \tag{24}$$

$$I(x,y) = \begin{cases} 0, & \text{if } y < 1, \\ 1, & \text{if } y = 1, \end{cases} \tag{25}$$

$$I(x,y) = t_2^{-1}(\min(c_x t_1(y), t_2(0)), \tag{26}$$

where $t_1, t_2 \colon [0,1] \to [0, \infty]$ are continuous, strictly decreasing functions with $t_1(1) = t_2(1) = 0$, $t_1(0) = \infty$ and $t_2(0) < \infty$.

Please note that these solutions are exactly the same as in the results already presented by Baczyński in [4, Theorem 5.1].

Theorem 5.3. *Let T_1 be a continuous t-norm, T_2 be a nilpotent t-norm, $I \colon [0,1]^2 \to [0,1]$ be a binary function and fix $x \in [0,1]$. If $\mu_{(T_1, I, x)} < \nu_{(T_1, I, x)}$ and the corresponding generating t-norm T_{α_0} of T_1 on the generating subinterval $[\mu_{(T_1, I, x)}, \nu_{(T_1, I, x)}] = [a_{\alpha_0}, b_{\alpha_0}]$ is nilpotent, then the following statements are equivalent:*

(i) The triple of functions $(T_1, T_2, I(x, \cdot))$ satisfies Eq. (3) for all $y, z \in [0, 1]$.

(ii) T_1 admits the representation (5), and there exist continuous, strictly decreasing functions $t_{\alpha_0}, t_2 \colon [0, 1] \to [0, \infty]$ with $t_{\alpha_0}(1) = t_2(1) = 0$, $t_{\alpha_0}(0) < \infty$, $t_2(0) < \infty$, which are uniquely determined up to positive multiplicative constants, such that the corresponding generating t-norm T_{α_0} of T_1 on the generating subinterval $[a_{\alpha_0}, b_{\alpha_0}]$ and T_2 admit the representation (4) with t_{α_0}, t_2, respectively, and for the mentioned above $x \in [0, 1]$, the vertical section $I(x, \cdot)$ has one of the following forms:

$$I(x, y) = \begin{cases} 0, & \text{if } y \in [0, b_{\alpha_0}), \\ 1, & \text{if } y \in [b_{\alpha_0}, 1], \end{cases} \tag{27}$$

$$I(x, y) = \begin{cases} 0, & \text{if } y \in [0, a_{\alpha_0}], \\ t_2^{-1}(\min(c_x t_{\alpha_0}(\frac{y - a_{\alpha_0}}{b_{\alpha_0} - a_{\alpha_0}}), t_2(0))), & \text{if } y \in [a_{\alpha_0}, b_{\alpha_0}], \\ 1, & \text{if } y \in [b_{\alpha_0}, 1], \end{cases} \tag{28}$$

with a certain $c_x \in [\frac{t_2(0)}{t_{\alpha_0}(0)}, \infty)$, uniquely determined up to a positive multiplicative constant depending on constants for t_{α_0} and t_2, i.e., if $t'_{\alpha_0}(y) = a t_{\alpha_0}(y)$, $t'_2(y) = b t_2(y)$ for all $y \in [0, 1]$ and some $a, b \in (0, \infty)$, and we assume that $t_2^{-1}(\min(c_x t_{\alpha_0}(\frac{y - a_{\alpha_0}}{b_{\alpha_0} - a_{\alpha_0}}), t_2(0))) = t_2'^{-1}(\min(c'_x t'_{\alpha_0}(\frac{y - a_{\alpha_0}}{b_{\alpha_0} - a_{\alpha_0}}), t'_2(0)))$, $y \in [a_{\alpha_0}, b_{\alpha_0}]$, then $c'_x = \frac{b}{a} c_x$.

Remark 5.4. In particular, when T_1, T_2 are nilpotent t-norms, then there are only two idempotent elements: 0 and 1. Therefore, for any fixed $x \in [0, 1]$ we get three possible cases:

- If $\mu_{(T_1, I, x)} = \nu_{(T_1, I, x)} = 0$, then the only solution is $I(x, y) = 0$ for all $y \in [0, 1]$ (it follows from Theorem 3.4).
- If $\mu_{(T_1, I, x)} = \nu_{(T_1, I, x)} = 1$, then the only solution is $I(x, y) = 1$ for all $y \in [0, 1]$ (it follows again from Theorem 3.4).
- If $\mu_{(T_1, I, x)} < \nu_{(T_1, I, x)}$, then $\mu_{(T_1, I, x)} = 0$ and $\nu_{(T_1, I, x)} = 1$ and from Theorem 5.3 we get two possible solutions:

$$I(x, y) = \begin{cases} 0, & \text{if } y < 1, \\ 1, & \text{if } y = 1, \end{cases} \tag{29}$$

$$I(x, y) = t_2^{-1}(\min(c_x t_1(y), t_2(0)), \tag{30}$$

where $t_1, t_2 \colon [0, 1] \to [0, \infty]$ are continuous, strictly decreasing functions with $t_1(1) = t_2(1) = 0$, $t_1(0) < \infty$ and $t_2(0) < \infty$.

Please note that these solutions are exactly the same as in the results already presented by Qin and Baczyński in [20, Theorem 2.2].

Next, let us consider continuous solutions of Eq. (3).

Theorem 5.5. *Let T_1 be a continuous t-norm, T_2 be a nilpotent t-norm and $I: [0,1]^2 \to [0,1]$ be a continuous binary function. Then the following statements are equivalent:*

(i) The triple of functions (T_1, T_2, I) satisfies Eq. (3) for all $x, y, z \in [0,1]$.

(ii) T_1 admits the representation (5), and there exists a continuous, strictly decreasing function $t_2: [0,1] \to [0,\infty]$ with $t_2(1) = 0$, $t_2(0) = \infty$, which is uniquely determined up to positive multiplicative constants, such that T_2 admits the representation (4) with t_2, I has the form either $I = 0$ or $I = 1$ or there exist two constants $a < b \in [0,1]$ such that $\mu_{(T_1, I, x)} = a, \nu_{(T_1, I, x)} = b$ for all $x \in [0,1]$, and there exists a continuous, strictly decreasing functions $t_a: [0,1] \to [0,\infty]$ with $t_a(1) = 0$, $t_a(0) = \infty$, which are uniquely determined up to positive multiplicative constants, such that the corresponding generating t-norm T_a of T_1 on the generating subinterval $[a,b]$ admits the representation (4) with t_a, and there exists a continuous function $c: [0,1] \to (0,\infty)$, uniquely determined up to a positive multiplicative constant depending on constants for t_a and t_2, such that, for $x, y \in [0,1]$, I has the form

$$I(x,y) = \begin{cases} 0, & \text{if } y \in [0,a], \\ t_2^{-1}(\min(c(x)t_a(\frac{y-a}{b-a}), t_2(0))), & \text{if } y \in [a,b], \\ 1, & \text{if } y \in [b,1], \end{cases} \tag{31}$$

or there exist two constants $a < b \in [0,1]$ such that $\mu_{(T_1, I, x)} = a, \nu_{(T_1, I, x)} = b$ for all $x \in [0,1]$, and there exists a continuous, strictly decreasing functions $t_a: [0,1] \to [0,\infty]$ with $t_a(1) = 0$, $t_a(0) < \infty$, which are uniquely determined up to positive multiplicative constants, such that the corresponding generating t-norm T_a of T_1 on the generating subinterval $[a,b]$ admits the representation (4) with t_a, and there exists a continuous function $c: [0,1] \to [\frac{t_2(0)}{t_a(0)}, \infty)$, uniquely determined up to a positive multiplicative constant depending on constants for t_a and t_2, such that I has the form in Eq. (31).

Corollary 5.6. *If T_1 is a continuous t-norm and T_2 is a nilpotent t-norm, then there are no continuous solutions of Eq. (3) which satisfy axiom I3.*

Proof. Let I be a continuous binary function satisfying axiom I3. By Theorem 5.5, I must have the form in Eq. (31). But in this case we get $I(0,0) = 0$ because of $a \geq 0$, which is a contradiction. □

From the above corollary it is obvious that we need to look for the solutions of Eq. (3), which are continuous except at the point $(0,0)$.

Theorem 5.7. *Let T_1 be a continuous t-norm, T_2 be a nilpotent t-norm and $I: [0,1]^2 \to [0,1]$ be a continuous fuzzy implication except at the vertical section $I(0,y) = 1$ for $y \in [0,a]$. Then the following statements are equivalent:*

(i) *The triple of functions* (T_1, T_2, I) *satisfies Eq.* (3) *for all* $x, y, z \in [0, 1]$.

(ii) T_1 *admits the representation* (5), *and there exist two constants* $a < b \in [0, 1]$ *such that* $\mu_{(T_1, I, x)} = a, \nu_{(T_1, I, x)} = b$ *for all* $x \in [0, 1]$, *and there exist continuous, strictly decreasing function* $t_a, t_2 \colon [0, 1] \to [0, \infty]$ *with* $t_a(1) = t_2(1) = 0$, $t_a(0) = \infty, t_2(0) < \infty$, *which is uniquely determined up to positive multiplicative constants, such that the corresponding generating t-norm* T_a *of* T_1 *on the generating subinterval* $[a, b]$ *and* T_2 *admit the representation* (4) *with* t_a *and* t_2 *respectively, and there exists a continuous and increasing function* $c \colon (0, 1] \to (0, \infty), c(0) = 0$, *uniquely determined up to a positive multiplicative constant depending on constants for* t_a *and* t_2, *such that* I *has the form*

$$I(x, y) = \begin{cases} 1, & \text{if } x = 0, y \in [0, a], \\ 0, & \text{if } x \neq 0, y \in [0, a], \\ t_2^{-1}(\min(c(x) t_a(\frac{y-a}{b-a}), t_2(0))), & \text{if } x \in [0, 1], y \in [a, b], \\ 1, & \text{if } x \in [0, 1], y \in [b, 1]. \end{cases} \tag{32}$$

6   Solutions to Eqs. (3) and (2) When $T_1 = T_M$ and T_2 Is a Continuous Archimedean T-norm

In Corollary 3.6 we have presented the solutions of Eq. (3) when $T_1 = T_M$ the minimum t-norm. Between all solutions we can easily find infinitely many solutions which are fuzzy implications. In this case for $x = 0$ we should have $c_0 = 0$ and the vertical section $I(x, \cdot)$ should have the form (10). Also I should be decreasing with respect to the second variable, so it is not possible that $c_{x_1} > c_{x_2}$ for $x_1 < x_2$. Otherwise let us take any $x_0 \in (c_{x_2}, c_{x_1})$ and independently to the solution (9) and (10) we get $I(x_1, x_0) = 0$ and $I(x_2, x_0) = 1$; a contradiction. We should also remember that $I(x, 1) = 1$ for every $x \in [0, 1]$, so it is not possible to have simultaneously the solution (9) and $c_x = 1$.

Now, between described above solutions it is not difficult to indicate solutions which, in addition, satisfy Eq. (2) with some strong negation N. This equation implies that, for any $x, y \in [0, 1]$, we should have either

$$y \leq c_x \iff N(x) \leq c_{N(y)} \quad \text{or} \quad y < c_x \iff N(x) < c_{N(y)}.$$

7   Solutions to Eqs. (3) and (2) When Both T_1, T_2 Are Continuous Archimedean T-norms

Characterizations of solutions (fuzzy implications) to the system of functional equations consisting of Eq. (3) and Eq. (2), when both t-norms T_1, T_2 are continuous Archimedean, have been presented recently by Qin and Baczyński in [20]. In this section we recall the main results presented in this article.

Theorem 7.1. *Let T_1, T_2 be strict t-norms and N be a strong negation. For a fuzzy implication I, which is continuous except at the points $(0,0)$ and $(1,1)$, the following statements are equivalent:*

(i) *The 4-tuple of functions T_1, T_2, I, N satisfies the system of functional equations (3) and (2) for all $x,y,z \in [0,1]$.*

(ii) *There exist continuous, strictly decreasing functions $t_1,t_2 \colon [0,1] \to [0,\infty]$ with $t_1(1) = t_2(1) = 0$, $t_1(0) = t_2(0) = \infty$, which are uniquely determined up to positive multiplicative constants, such that T_1, T_2 admit the representation (4) with t_1, t_2, respectively, an increasing bijection $\varphi \colon [0,1] \to [0,1]$ such that N admits the representation (6), and a constant $r \in]0,\infty[$, such that I has the following form*

$$I(x,y) = \begin{cases} 1, & \text{if } x = y = 0 \text{ or } x = y = 1, \\ t_2^{-1}(rt_1(\varphi^{-1}(1 - \varphi(x)))t_1(y)), & \text{otherwise,} \end{cases}$$

for all $x,y \in [0,1]$.

Theorem 7.2. *Let T_1, T_2 be nilpotent t-norms and N be a strong negation. For a fuzzy implication I the following statements are equivalent:*

(i) *The 4-tuple of functions T_1, T_2, I, N satisfies the system of functional equations (3) and (2) for all $x,y,z \in [0,1]$.*

(ii) *I is the least fuzzy implication (see [6]):*

$$I_0(x,y) = \begin{cases} 1, & \text{if } x = 0 \text{ or } y = 1, \\ 0, & \text{otherwise,} \end{cases} \qquad x,y \in [0,1].$$

Theorem 7.3. *Let T_1 be a strict t-norm, T_2 be a nilpotent t-norm and N be a strong negation. For a fuzzy implication I, which is continuous except at the points $(0,0)$ and $(1,1)$, the following statements are equivalent:*

(i) *The 4-tuple of functions T_1, T_2, I, N satisfies the system of functional equations (3) and (2) for all $x,y,z \in [0,1]$.*

(ii) *There exist continuous and strictly increasing functions $t_1,t_2 \colon [0,1] \to [0,\infty]$ with $t_1(1) = t_2(1) = 0$, $t_1(0) = \infty$ and $t_2(0) < \infty$, which are uniquely determined up to positive multiplicative constants, such that T_1, T_2 admit the representation (4) with t_1,t_2, respectively, an increasing bijection $\varphi \colon [0,1] \to [0,1]$ such that N admits the representation (6), and a constant $r \in]0,\infty[$ such that I has the following form*

$$I(x,y) = \begin{cases} 1, & \text{if } x = y = 0 \text{ or } x = y = 1, \\ t_2^{-1}\left(\min(rt_1(\varphi^{-1}(1 - \varphi(x)))t_1(y), t_2(0))\right), & \text{otherwise,} \end{cases}$$

for all $x,y \in [0,1]$.

Theorem 7.4. *Let T_1 be a nilpotent t-norm, T_2 be a strict t-norm and N be a strong negation. For a fuzzy implication I the following statements are equivalent:*

(i) *The 4-tuple of functions T_1, T_2, I, N satisfies the system of functional equations (3) and (2) for all $x, y, z \in [0, 1]$.*
(ii) *I is the least fuzzy implication I_0.*

Therefore, applying the method used by Baczyński in [2], we see that the system of functional equations consisting of Eq. (2) and Eq. (3) has many solutions when T_1 is a strict t-norm, T_2 is a continuous Archimedean t-norm, N is a strong negation and I is a continuous binary function except at the points $(0, 0)$ and $(1, 1)$.

8 Solutions to Eqs. (3) and (2) When T_1 Is a Continuous Non-Archimedean, Non-idempotent T-norm and T_2 Is a Strict T-norm

In this section, we characterize solutions to the system of functional equations consisting of Eq. (2) and Eq. (3) when T_1 is a continuous non-Archimedean t-norm (and different from the minimum) and T_2 is a strict t-norm. Our presentation in this and the next section is based on facts presented by us in [19].

From Remark 4.7 we can draw a conclusion that if I is not continuous at point $(0, 0)$, then I is also not continuous on the vertical section $I(0, y)$ for all $y \in [0, a]$. From the contrapositive symmetry of implication Eq. (2) we get that I is also not continuous on the horizontal section $I(x, 1)$ for all $x \in [N(a), 1]$, where N is a strong negation. Thus, we get the following result.

Theorem 8.1. *Let T_1 be a continuous, non-Archimedean and non-idempotent t-norm, T_2 be a strict t-norm, N be a strong negation and $I: [0, 1]^2 \to [0, 1]$ be a continuous binary function except at the points $(0, 0)$ and $(1, 1)$, which satisfies axiom I3. Then the 4-tuple of functions T_1, T_2, I, N does not satisfy the system of functional equations consisting of Eq. (2) and Eq. (3).*

Theorem 8.2. *Let T_1 be a continuous, non-Archimedean and non-idempotent t-norm, T_2 be a strict t-norm, N be a strong negation and $I: [0, 1]^2 \to [0, 1]$ be a binary function. Further, let us assume that there exist two constants $a < b \in (0, 1]$ such that $\mu_{(T_1, I, x)} = a$, $\nu_{(T_1, I, x)} = b$ for all $x \in [0, 1]$, and two continuous, strictly decreasing functions $t_a, t_2: [0, 1] \to [0, \infty]$ with $t_a(1) = t_2(1) = 0$, $t_a(0) = t_2(0) = \infty$, which are uniquely determined up to positive multiplicative constants, such that the corresponding generating t-norm T_a of T_1 on the generating subinterval $[a, b]$ and T_2 admit the representation (4) with t_a and t_2, respectively. If I satisfies axiom I3 and is continuous except at the vertical section $I(0, y) = 1$ for $y \in [0, a]$ and the horizontal section $I(x, 1)$ for $x \in [N(a), 1]$, then the following statements are equivalent:*

(i) *The 4-tuple of functions T_1, T_2, I, N satisfies the system of functional equations (3) and (2) for all $x, y, z \in [0, 1]$.*
(ii) *$b = 1$ and there exists a constant $r \in (0, \infty)$ such that I has the following form*

$$I(x,y) = \begin{cases} 1, & \text{if } x = 0 \text{ or } y = 1, \\ t_2^{-1}(r \cdot t_a(\frac{N(x)-a}{1-a}) \cdot t_a(\frac{y-a}{1-a})), & \text{if } x \in (0, N(a)) \text{ and } y \in (a, 1), \\ 0, & \text{otherwise.} \end{cases}$$

Theorem 8.3. *Let T_1 be a continuous non-Archimedean and non-idempotent t-norm, T_2 be a strict t-norm, N be a strong negation, $I \colon [0,1]^2 \to [0,1]$ a continuous binary function except at the vertical section $I(0,y) = 1$ for $y \in [0,1]$ and the horizontal section $I(x,1)$ for $x \in [0,1]$, which satisfies axiom I3. Then the following statements are equivalent:*

(i) The 4-tuple of functions T_1, T_2, I, N satisfies the system of functional equations (3) and (2) for all $x,y,z \in [0,1]$.
(ii) I is the least fuzzy implication I_0.

9 Solutions to Eqs. (3) and (2) When T_1 Is a Continuous Non-Archimedean and Non-idempotent T-norm and T_2 Is a Nilpotent T-norm

In this section, we characterize all solutions to the system of functional equations consisting of Eq. (2) and Eq. (3) when T_1 is a continuous non-Archimedean t-norm (and different from the minimum) and T_2 is a nilpotent t-norm. The results are based on facts included in [19]. Similar to the analysis in previous section, it is enough to consider solutions I which are not continuous on the vertical section $I(0,y)$ for all $y \in [0,a]$ and on the horizontal section $I(x,1)$ for all $x \in [N(a),1]$. At first, we consider the situation when the corresponding generating t-norm T_a of T_1 on the generating subinterval $[a,b]$ is strict.

Theorem 9.1. *Let T_1 be a continuous non-Archimedean and non-idempotent t-norm, T_2 be a nilpotent t-norm, N be a strong negation, $I \colon [0,1]^2 \to [0,1]$ be a continuous binary function except at the points $(0,0)$ and $(1,1)$, which satisfies aziom I3. Let us assume that there exist one constant $b \in (0,1]$ such that $\mu_{(T_1,I,x)} = 0$, $\nu_{(T_1,I,x)} = b$ for all $x \in [0,1]$, and two continuous, strictly decreasing function t_0, $t_2 \colon [0,1] \to [0,\infty]$ with $t_0(1) = t_2(1) = 0$, $t_0(0) = \infty$, $t_2(0) < \infty$, which are uniquely determined up to positive multiplicative constants, such that the corresponding generating t-norm T_0 of T_1 on the generating subinterval $[0,b]$ and T_2 admit the representation (4) with t_0 and t_2, respectively. Then the 4-tuple of functions T_1,T_2,I,N does not satisfy Eq. (3) and Eq. (2).*

Theorem 9.2. *Let T_1 be a continuous non-Archimedean and non-idempotent t-norm, T_2 be a nilpotent t-norm, N be a strong negation and $I \colon [0,1]^2 \to [0,1]$ be a binary function. Let us assume that there exist two constants $a,b \in (0,1]$, $a < b$ such that $\mu_{(T_1,I,x)} = a$, $\nu_{(T_1,I,x)} = b$ for all $x \in [0,1]$, and two continuous, strictly decreasing function $t_a, t_2 \colon [0,1] \to [0,\infty]$ with $t_a(1) = t_2(1) = 0$, $t_a(0) = \infty$, $t_2(0) < \infty$, which are uniquely determined up to positive multiplicative constants, such that the corresponding generating t-norm T_a of T_1 on the generating subinterval $[a,b]$ and T_2*

admit the representation (4) *with* t_a *and* t_2, *respectively. If I satisfies axiom I3 and is continuous except at the vertical section* $I(0,y) = 1$ *for* $y \in [0,a]$ *and the horizontal section* $I(x,1)$ *for* $x \in [N(a),1]$, *then the following statements are equivalent:*

(i) *The 4-tuple of functions* T_1, T_2, I, N *satisfies the system of functional equations* (3) *and* (2) *for all* $x,y,z \in [0,1]$.

(ii) $b = 1$ *and there exists a constant* $r \in (0,\infty)$ *such that* I *has the following form*

$$I(x,y) = \begin{cases} 1, & \text{if } x = 0 \text{ or } y = 1, \\ t_2^{-1}(\min(r \cdot t_a(\frac{N(x)-a}{1-a}) \cdot t_a(\frac{y-a}{1-a}), t_2(0))), & \text{if } x \in (0,N(a)) \text{ and } y \in (a,1), \\ 0, & \text{otherwise.} \end{cases}$$

Now, let us consider the case where the generating t-norm T_a of T_1 on the generating subinterval $[a,b]$ is nilpotent.

Theorem 9.3. *Let* T_1 *be a continuous non-Archimedean and non-idempotent t-norm,* T_2 *be a nilpotent t-norm,* N *be a strong negation,* $I\colon [0,1]^2 \to [0,1]$ *be a continuous binary function except at the points* $(0,0)$ *and* $(1,1)$, *which satisfies axiom I3. Let us assume that there exist one constant* $b \in (0,1]$ *such that* $\mu_{(T_1,I,x)} = 0$, $\nu_{(T_1,I,x)} = b$ *for all* $x \in [0,1]$, *and two continuous, strictly decreasing function* $t_0, t_2 \colon [0,1] \to [0,\infty]$ *with* $t_0(1) = t_2(1) = 0$, $t_0(0) < \infty$, $t_2(0) < \infty$, *which are uniquely determined up to positive multiplicative constants, such that the corresponding generating t-norm* T_0 *of* T_1 *on the generating subinterval* $[0,b]$ *and* T_2 *admit the representation* (4) *with* t_0 *and* t_2, *respectively. Then the 4-tuple of functions* T_1, T_2, I, N *does not satisfy Eq.* (3) *and Eq.* (2).

Theorem 9.4. *Let* T_1 *be a continuous non-Archimedean and non-idempotent t-norm,* T_2 *be a nilpotent t-norm,* N *be a strong negation and* $I\colon [0,1]^2 \to [0,1]$ *be a binary function which satisfies axiom I3. Let us assume that there exist two constants* $a,b \in (0,1]$, $a < b$ *such that* $\mu_{(T_1,I,x)} = a$, $\nu_{(T_1,I,x)} = b$ *for all* $x \in [0,1]$, *and two continuous, strictly decreasing function* $t_a, t_2 \colon [0,1] \to [0,\infty]$ *with* $t_a(1) = t_2(1) = 0$, $t_a(0) < \infty$, $t_2(0) < \infty$, *which are uniquely determined up to positive multiplicative constants, such that the corresponding generating t-norm* T_a *of* T_1 *on the generating subinterval* $[a,b]$ *and* T_2 *admit the representation* (4) *with* t_a *and* t_2, *respectively. If*

* *I is continuous binary function except at the points* $(0,0)$ *and* $(1,1)$

or

* *I is continuous except at the vertical section* $I(0,y) = 1$ *for* $y \in [0,a]$ *and the horizontal section* $I(x,1)$ *for* $x \in [N(a),1]$,

then the 4-tuple of functions T_1, T_2, I, N *does not satisfy Eq.* (3) *and Eq.* (2).

Finally, we have the following result.

Theorem 9.5. *Let* T_1 *be a continuous non-Archimedean and non-idempotent t-norm,* T_2 *be a nilpotent t-norm,* N *be a strong negation,* $I\colon [0,1]^2 \to [0,1]$ *a continuous binary function except at the vertical section* $I(0,y) = 1$ *for* $y \in [0,1]$ *and* *the horizontal section $I(x, 1)$ for $x \in [0, 1]$, which satisfies axiom I3. Then the following statements are equivalent:*

(i) The 4-tuple of functions T_1, T_2, I, N satisfies the system of functional equations (3) and (2) for all $x, y, z \in [0, 1]$.
(ii) I is the least fuzzy implication I_0.

10 Conclusion

In this work, we summarize the sufficient and necessary conditions of solutions for Eq. (3) and the system of functional equations consisting of Eq. (2) and Eq. (3), when T_1 is a continuous t-norm, T_2 is a continuous Archimedean t-norm, N is a strong negation and I is an unknown function (in particular fuzzy implication). We also underline that our method can be applied to the three other distributivity functional equations for fuzzy implications.

Acknowledgements. The first author, Feng Qin, was supported by National Natural Science Foundation of China (No.60904041), Jiangxi Natural Science Foundation (No.2009GQS0055) and Scientific Research Foundation of Jiangxi Provincial Education Department(No.GJJ08160). Tel: +86 791 3863755. The second author, M. Baczyński, was supported by the Polish Ministry of Science and Higher Education Grant Nr N N519 384936.

References

1. Baczyński, M.: On a class of distributive fuzzy implications. Internat. J. Uncertainty, Fuzziness and Knowledge-Based Systems 9, 229–238 (2001)
2. Baczyński, M.: Contrapositive symmetry of distributive fuzzy implications. Internat. Uncertainty, Fuzziness and Knowledge-Based Systems 10, 135–147 (2002)
3. Baczyński, M.: On the distributivity of fuzzy implications over continuous and Archimedean triangular conorms. Fuzzy Sets and Systems 161, 1406–1419 (2010)
4. Baczyński, M.: On the Distributivity of Fuzzy Implications over Continuous Archimedean Triangular Norms. In: Rutkowski, L., Scherer, R., Tadeusiewicz, R., Zadeh, L.A., Zurada, J.M. (eds.) ICAISC 2010, Part I. LNCS (LNAI), vol. 6113, pp. 3–10. Springer, Heidelberg (2010)
5. Baczyński, M., Drewniak, J.: Conjugacy Classes of Fuzzy Implication. In: Reusch, B. (ed.) Fuzzy Days 1999. LNCS, vol. 1625, pp. 287–298. Springer, Heidelberg (1999)
6. Baczyński, M., Jayaram, B.: Fuzzy implications. STUDFUZZ, vol. 231. Springer, Berlin (2008)
7. Baczyński, M., Jayaram, B.: On the distributivity of fuzzy implications over nilpotent or strict triangular conorms. IEEE Trans. Fuzzy Syst. 17, 590–603 (2009)
8. Balasubramaniam, J., Rao, C.J.M.: On the distributivity of implication operators over T- and S-norms. IEEE Trans. Fuzzy Syst. 12, 194–198 (2004)
9. Bustince, H., Burillo, P., Soria, F.: Automorphisms, negations and implication operators. Fuzzy Sets and Systems 134, 209–229 (2003)
10. Combs, W.E., Andrews, J.E.: Combinatorial rule explosion eliminated by a fuzzy rule configuration. IEEE Trans. Fuzzy Syst. 6, 1–11 (1998)
11. Combs, W.E.: Author's reply. IEEE Trans. Fuzzy Syst. 7, 371, 478–479 (1999)

12. Dick, S., Kandel, A.: Comments on "Combinatorial rule explosion eliminated by a fuzzy rule configuration". IEEE Trans. Fuzzy Syst. 7, 475–477 (1999)
13. Fodor, J., Roubens, M.: Fuzzy Preference Modelling and Multicriteria Decision Support. Kluwer, Dordrecht (1994)
14. Gottwald, S.: A Treatise on Many-Valued Logics. Research Studies, Baldock (2001)
15. Klement, E.P., Mesiar, R., Pap, E.: Triangular Norms. Kluwer, Dordrecht (2000)
16. Kuczma, M.: An Introduction to the Theory of Functional Equations and Inequalities: Cauchy's Equation and Jensen's Inequality. PWN-Polish Scientific Publishers & Silesian University, Warszawa-Kraków-Katowice (1985)
17. Ling, C.H.: Representation of associative functions. Publ. Math. Debrecen 12, 189–212 (1965)
18. Mendel, J.M., Liang, Q.: Comments on "Combinatorial rule explosion eliminated by a fuzzy rule configuration". IEEE Trans. Fuzzy Syst. 7, 369–371 (1999)
19. Qin, F., Baczyński, M.: On distributivity equations of implications and contrapositive symmetry equations of implications. Fuzzy Sets and Systems (submitted)
20. Qin, F., Baczyński, M.: Some remarks on the distributivity equation and the contrapositive symmetry for fuzzy implications – the continuous, Archimedean t-norms case. International Journal of Approximate Reasoning (submitted)
21. Qin, F., Baczyński, M., Xie, A.: Distributive equations of implications based on continuous triangular norms (I). IEEE Trans. Fuzzy Systems 20, 153–167 (2012)
22. Qin, F., Yang, L.: Distributive equations of implications based on nilpotent triangular norms. Internat. J. Approx. Reason. 51, 984–992 (2010)
23. Qin, F., Zhao, B.: The distributive equations for idempotent uninorms and nullnorms. Fuzzy Sets and Systems 155, 446–458 (2005)
24. Ruiz-Aguilera, D., Torrens, J.: Distributivity of strong implications over conjunctive and disjunctive uninorms. Kybernetika 42, 319–336 (2005)
25. Ruiz-Aguilera, D., Torrens, J.: Distributivity of residual implications over conjunctive and disjunctive uninorms. Fuzzy Sets and Systems 158, 23–37 (2007)
26. Trillas, E.: Sobre funciones de negación en la teoria de conjuntos difusos. Stochastica 3, 47–59, (1979); english translation In: Barro, S., Bugarin, A., Sobrino, A. (eds.), Advances in fuzzy logic. Public. Univ. Santiago de Compostela, pp. 31–45 (1998)
27. Trillas, E., Alsina, C.: On the law $[p \wedge q \to r] = [(p \to r) \vee (q \to r)]$ in fuzzy logic. IEEE Trans. Fuzzy Syst. 10, 84–88 (2002)
28. Trillas, E., Mas, M., Monserrat, M., Torrens, J.: On the representation of fuzzy rules. Internat. J. Approx. Reason. 48, 583–597 (2008)
29. Türksen, I.B., Kreinovich, V., Yager, R.R.: A new class of fuzzy implications. Axioms of fuzzy implication revisted. Fuzzy Sets and Systems 100, 267–272 (1998)
30. Yang, L., Qin, F.: Distributive equations based on fuzzy implications. In: IEEE International Conference on Fuzzy Systems, Korea, pp. 560–563 (2009)

Implication Functions in Interval-Valued Fuzzy Set Theory

Glad Deschrijver

Abstract. Interval-valued fuzzy set theory is an extension of fuzzy set theory in which the real, but unknown, membership degree is approximated by a closed interval of possible membership degrees. Since implications on the unit interval play an important role in fuzzy set theory, several authors have extended this notion to interval-valued fuzzy set theory. This chapter gives an overview of the results pertaining to implications in interval-valued fuzzy set theory. In particular, we describe several possibilities to represent such implications using implications on the unit interval, we give a characterization of the implications in interval-valued fuzzy set theory which satisfy the Smets-Magrez axioms, we discuss the solutions of a particular distributivity equation involving strict t-norms, we extend monoidal logic to the interval-valued fuzzy case and we give a soundness and completeness theorem which is similar to the one existing for monoidal logic, and finally we discuss some other constructions of implications in interval-valued fuzzy set theory.

1 Introduction

Fuzzy set theory has been introduced by Zadeh [57] in order to deal with the imprecision, ignorance and vagueness present in the real world, and has been applied successfully in several areas. In fuzzy set theory the membership of an object in a set is determined by assigning a real number between 0 and 1, called the membership degree of the object in the set. However, in some real problems, it is very difficult to determine a correct value (if there is one) for the membership degrees. In many cases only an approximated value of the membership degree is given. This kind of uncertainty in the membership degrees has motivated several extensions of Zadeh's fuzzy set theory, such as Atanassov's intuitionistic fuzzy set theory [1],

Glad Deschrijver
Fuzziness and Uncertainty Modelling Research Unit, Department of Applied Mathematics and Computer Science, Ghent University, Krijgslaan 281(S9), B–9000 Gent, Belgium
e-mail: Glad.Deschrijver@UGent.be

M. Baczyński et al. (Eds.): *Adv. in Fuzzy Implication Functions*, STUDFUZZ 300, pp. 73–99.
DOI: 10.1007/978-3-642-35677-3_4   © Springer-Verlag Berlin Heidelberg 2013

interval-valued fuzzy set theory [43], type-2 fuzzy set theory [58], ... Interval-valued fuzzy sets assign to each object instead of a single number a closed interval which approximates the real, but unknown, membership degree. As such, interval-valued fuzzy set theory forms a good balance between the ease of use of fuzzy set theory and the expressiveness of type-2 fuzzy set theory. Since the underlying lattice of Atanassov's intuitionistic fuzzy set theory is isomorphic to the underlying lattice of interval-valued fuzzy set theory, any results about any functions on any of those lattices hold for both theories. Therefore we will focus in this work to functions defined on the underlying lattice of interval-valued fuzzy set theory. Interval-valued fuzzy sets and Atanassov's intuitionistic fuzzy sets have been investigated both theoretically and practically by many researchers [12, 13, 15, 33, 38, 40, 45, 53, 54, 55, 60].

Since implications on the unit interval play an important role in fuzzy set theory [9], several authors have extended this notion to interval-valued fuzzy set theory [4, 6, 7, 8, 11, 17, 26, 48]. This chapter gives an overview of the results pertaining to implications in interval-valued fuzzy set theory. In the next section we start with some preliminary definitions concerning the underlying structure of interval-valued fuzzy set theory and some functions which we will need later on. This section is followed by several sections in which we give an overview of known results.

2 Preliminary Definitions

2.1 The Lattice $\mathscr{L}^I$

The underlying lattice $\mathscr{L}^I$ of interval-valued fuzzy set theory is given as follows.

Definition 1. We define $\mathscr{L}^I = (L^I, \leq_{L^I})$, where

$$L^I = \{[x_1, x_2] \mid (x_1, x_2) \in [0,1]^2 \text{ and } x_1 \leq x_2\},$$
$$[x_1, x_2] \leq_{L^I} [y_1, y_2] \iff (x_1 \leq y_1 \text{ and } x_2 \leq y_2), \text{ for all } [x_1, x_2], [y_1, y_2] \text{ in } L^I.$$

Similarly as Lemma 2.1 in [24] it is shown that $\mathscr{L}^I$ is a complete lattice.

Definition 2. [34, 43] An interval-valued fuzzy set on U is a mapping $A : U \to L^I$.

Definition 3. [1, 2, 3] An intuitionistic fuzzy set in the sense of Atanassov on U is a set

$$A = \{(u, \mu_A(u), \nu_A(u)) \mid u \in U\},$$

where $\mu_A(u) \in [0,1]$ denotes the membership degree and $\nu_A(u) \in [0,1]$ the non-membership degree of u in A and where for all $u \in U$, $\mu_A(u) + \nu_A(u) \leq 1$.

An intuitionistic fuzzy set in the sense of Atanassov A on U can be represented by the $\mathscr{L}^I$-fuzzy set A given by

$$A : U \rightarrow L^I :$$
$$u \mapsto [\mu_A(u), 1 - \nu_A(u)], \quad \forall u \in U.$$

In Figure 1 the set L^I is shown. Note that to any element $x = [x_1, x_2]$ of L^I there corresponds a point $(x_1, x_2) \in \mathbb{R}^2$.

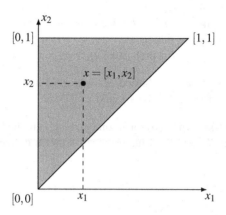

Fig. 1 The grey area is L^I

In the sequel, if $x \in L^I$, then we denote its bounds by $x_1 = \mathrm{pr}_1(x)$ and $x_2 = \mathrm{pr}_2(x)$, i.e. $x = [x_1, x_2]$. The smallest and the largest element of $\mathscr{L}^I$ are given by $0_{\mathscr{L}^I} = [0, 0]$ and $1_{\mathscr{L}^I} = [1, 1]$. The hypothenuse of the triangle corresponds to the set $D = \{[x_1, x_1] \mid x_1 \in [0, 1]\}$ of values in L^I about which there is no indeterminacy and can be identified with the unit interval $[0, 1]$ from (classical) fuzzy set theory. The elements of D are called the *exact elements* of the lattice $\mathscr{L}^I$. Note that, for x, y in L^I, $x <_{L^I} y$ is equivalent to "$x \leq_{L^I} y$ and $x \neq y$", i.e. either $x_1 < y_1$ and $x_2 \leq y_2$, or $x_1 \leq y_1$ and $x_2 < y_2$. We denote by $x \ll_{L^I} y$: $x_1 < y_1$ and $x_2 < y_2$.

Bedregal et al. [12, 44] introduced the notion of *interval representation*, where an interval function $F : L^I \rightarrow L^I$ represents a real function $f : [0, 1] \rightarrow [0, 1]$ if for each $X \in L^I$, $f(x) \in F(X)$ whenever $x \in X$ (the interval X represents the real x). So, F is an interval representation of f if $F(X)$ includes all possible situations that could occur if the uncertainty in X were to be expelled. For $f : [0, 1] \rightarrow [0, 1]$, the function $\hat{f} : L^I \rightarrow L^I$ defined by

$$\hat{f}(X) = [\inf\{f(x) \mid x \in X\}, \sup\{f(x) \mid x \in X\}]$$

is an interval representation of f [12, 44]. Clearly, if F is also an interval representation of f, then for each $X \in L^I$, $\hat{f}(X) \subseteq F(X)$. Thus, $\hat{f}$ returns a narrower interval than any other interval representation of f and is therefore its *best interval representation*.

2.2 Triangular Norms, Implications and Negations on $\mathscr{L}^I$

Implications are often generated from other connectives. In this section we will introduce some of these connectives and give the construction of implications derived from these functions.

Definition 4. A t-norm on a complete lattice $\mathscr{L} = (L, \leq_L)$ is a commutative, associative, increasing mapping $\mathscr{T} : L^2 \to L$ which satisfies $\mathscr{T}(1_{\mathscr{L}}, x) = x$, for all $x \in L$.

A t-conorm on a complete lattice $\mathscr{L} = (L, \leq_L)$ is a commutative, associative, increasing mapping $\mathscr{S} : L^2 \to L$ which satisfies $\mathscr{S}(0_{\mathscr{L}}, x) = x$, for all $x \in L$.

Let $\mathscr{T}$ be a t-norm on a complete lattice $\mathscr{L} = (L, \leq_L)$ and $x \in L$, then we denote $x^{(n)_{\mathscr{T}}} = \mathscr{T}(x, x^{(n-1)_{\mathscr{T}}})$, for $n \in \mathbb{N} \setminus \{0, 1\}$, and $x^{(1)_{\mathscr{T}}} = x$.

Example 1. Some well-known t-(co)norms on $([0,1], \leq)$ are the Łukasiewicz t-norm T_L, the product t-norm T_P and the Łukasiewicz t-conorm defined by, for all x, y in $[0,1]$,

$$T_L(x,y) = \max(0, x+y-1),$$
$$T_P(x,y) = xy,$$
$$S_L(x,y) = \min(1, x+y).$$

For t-norms on $\mathscr{L}^I$, we consider the following special classes.

Lemma 1. *[21]*

- *Given t-norms T_1 and T_2 on $([0,1], \leq)$ with $T_1 \leq T_2$, the mapping $\mathscr{T}_{T_1,T_2} : (L^I)^2 \to L^I$ defined by, for all x, y in L^I,*

$$\mathscr{T}_{T_1,T_2}(x,y) = [T_1(x_1,y_1), T_2(x_2,y_2)].$$

  is a t-norm on $\mathscr{L}^I$.
- *Given a t-norm T on $([0,1], \leq)$, the mappings $\mathscr{T}_T : (L^I)^2 \to L^I$ and $\mathscr{T}'_T : (L^I)^2 \to L^I$ defined by, for all x, y in L^I,*

$$\mathscr{T}_T(x,y) = [T(x_1,y_1), \max(T(x_1,y_2), T(x_2,y_1))],$$
$$\mathscr{T}'_T(x,y) = [\min(T(x_1,y_2), T(x_2,y_1)), T(x_2,y_2)],$$

  are t-norms on $\mathscr{L}^I$.

Definition 5. [21] Let T_1, T_2 and T be t-norms on $([0,1], \leq)$. The t-norms $\mathscr{T}_{T_1,T_2}$, $\mathscr{T}_T$ and $\mathscr{T}'_T$ defined in Lemma 1 are called the t-representable t-norm on $\mathscr{L}^I$ with representatives T_1 and T_2, the optimistic t-norm and the pessimistic t-norm on $\mathscr{L}^I$ with representative T, respectively. In a similar way t-representable, pessimistic and optimistic t-conorms on $\mathscr{L}^I$ can be defined.

Note that $\mathscr{T}_{T,T}$ is the best interval representation of T. Furthermore, if T is continuous[1],

$$\mathscr{T}_{T,T}([x_1,x_2],[y_1,y_2]) = \{T(\alpha,\beta) \mid \alpha \in [x_1,x_2] \text{ and } \beta \in [y_1,y_2]\}. \tag{1}$$

Looking at the structure of $\mathscr{T}_T$, this t-norm has the same lower bound as the t-representable t-norm $\mathscr{T}_{T,T}$, but differs from it by its upper bound: instead of taking the "optimum" value $T(x_2,y_2)$, the second component is obtained by taking the maximum of $T(x_1,y_2)$ and $T(x_2,y_1)$. Hence it is not guaranteed that the interval $\mathscr{T}_T(x,y)$ contains all possible values $T(\alpha,\beta)$ for $\alpha \in [x_1,x_2]$ and $\beta \in [y_1,y_2]$. Rather (for continuous T),

$$\mathscr{T}_T([x_1,x_2],[y_1,y_2]) = \{T(\alpha,y_1) \mid \alpha \in [x_1,x_2]\} \cup \{T(x_1,\beta) \mid \beta \in [y_1,y_2]\}. \tag{2}$$

What this representation enforces is that, in eliminating the uncertainty from x and y, we have to impose for at least one of them the "worst" possible value (x_1, resp. y_1). Therefore, this could be called a *pessimistic* approach to the definition of a t-norm on $\mathscr{L}^I$, hence the name "pessimistic t-norm". Similarly, the adapted upper bound of $\mathscr{T}_T'$ reflects an optimistic approach.

A class of t-norms generalizing both the t-representable t-norms and the pessimistic t-norms can be introduced. Let T be a t-norm on $([0,1],\leq)$, and $t \in [0,1]$. Then the mapping $\mathscr{T}_{T,t} : (L^I)^2 \to L^I$ defined by, for all x,y in L^I,

$$\mathscr{T}_{T,t}(x,y) = [T(x_1,y_1), \max(T(t,T(x_2,y_2)), T(x_1,y_2), T(x_2,y_1))],$$

is a t-norm on $\mathscr{L}^I$ [23]. The usage of this class is that it allows the user to define $\mathscr{T}([0,1],[0,1]) = [0,t]$ arbitrarily. This can be useful in applications where in some situations one needs to impose that the conjunction of two completely unknown propositions is also unknown (e.g. "the sun will shine tomorrow" and "this night it will freeze"), while in other situations it would be more appropriate that the conjunction of two unknown statements is false (e.g. "this night it will freeze" and "this night it will be hot"). If $t = 0$, then we obtain the pessimistic t-norms, if $t = 1$, then we find t-representable t-norms. Clearly, since the lower bound of $\mathscr{T}_{T,t}(x,y)$ is independent of x_2 and y_2, the optimistic t-norms do not belong to this class as soon as $T \neq \min$.

Definition 6. An implication on a complete lattice $\mathscr{L} = (L,\leq_L)$ is a mapping $\mathscr{I} : L^2 \to L$ that is decreasing (w.r.t. $\leq_L$) in its first, and increasing (w.r.t. $\leq_L$) in its second argument, and that satisfies

$$\mathscr{I}(0_{\mathscr{L}},0_{\mathscr{L}}) = 1_{\mathscr{L}}, \quad \mathscr{I}(0_{\mathscr{L}},1_{\mathscr{L}}) = 1_{\mathscr{L}},$$
$$\mathscr{I}(1_{\mathscr{L}},1_{\mathscr{L}}) = 1_{\mathscr{L}}, \quad \mathscr{I}(1_{\mathscr{L}},0_{\mathscr{L}}) = 0_{\mathscr{L}}.$$

---

[1] The continuity is necessary in order to have an equality in (1). In the general case it only holds that the left-hand side of (1) is a subset of the right-hand side.

Definition 7. A negation on a complete lattice $\mathscr{L} = (L, \leq_L)$ is a decreasing mapping $\mathscr{N} : L \rightarrow L$ for which $\mathscr{N}(0_{\mathscr{L}}) = 1_{\mathscr{L}}$ and $\mathscr{N}(1_{\mathscr{L}}) = 0_{\mathscr{L}}$. If $\mathscr{N}(\mathscr{N}(x)) = x$, for all $x \in L$, then $\mathscr{N}$ is called involutive.

Proposition 1. *[6] Let IVFI be the set of all implications on $\mathscr{L}^I$. Then $(IVFI, \inf, \sup)$ is a complete lattice, i.e.*

$$(\forall t \in T)(\mathscr{I}_t \in IVFI) \implies (\sup_{t \in T} \mathscr{I}_t, \inf_{t \in T} \mathscr{I}_t) \in IVFI^2.$$

Corollary 1. *[6] $(IVFI, \inf, \sup)$ has the greatest element*

$$\mathscr{I}_1(x,y) = \begin{cases} 0_{\mathscr{L}^I}, & \text{if } x = 1_{\mathscr{L}^I} \text{ and } y = 0_{\mathscr{L}^I}, \\ 1_{\mathscr{L}^I}, & \text{otherwise}, \end{cases}$$

and the least element

$$\mathscr{I}_0(x,y) = \begin{cases} 1_{\mathscr{L}^I}, & \text{if } x = 0_{\mathscr{L}^I} \text{ or } y = 1_{\mathscr{L}^I}, \\ 0_{\mathscr{L}^I}, & \text{otherwise}. \end{cases}$$

Implications are often derived from other types of connectives. For our purposes, we consider S- and R-implications:

- let $\mathscr{T}$ be a t-norm on $\mathscr{L}$, then the residual implication or R-implication $\mathscr{I}_{\mathscr{T}}$ is defined by, for all x, y in L,

$$\mathscr{I}_{\mathscr{T}}(x,y) = \sup\{z \mid z \in L \text{ and } \mathscr{T}(x,z) \leq_L y\}; \qquad (3)$$

- let $\mathscr{S}$ be a t-conorm and $\mathscr{N}$ a negation on $\mathscr{L}$, then the S-implication $\mathscr{I}_{\mathscr{S},\mathscr{N}}$ is defined by, for all x, y in L,

$$\mathscr{I}_{\mathscr{S},\mathscr{N}}(x,y) = \mathscr{S}(\mathscr{N}(x),y). \qquad (4)$$

We say that a t-norm $\mathscr{T}$ on $\mathscr{L}$ satisfies the *residuation principle* if and only if, for all x, y, z in L,

$$\mathscr{T}(x,y) \leq_L z \iff y \leq_L \mathscr{I}_{\mathscr{T}}(x,z).$$

Example 2. The residual implications of the t-norms given in Example 1 are given by, for all x, y in $[0,1]$,

$$I_{T_L}(x,y) = \min(1, y + 1 - x),$$
$$I_{T_P}(x,y) = \min\left(1, \frac{y}{x}\right),$$

using the convention $\frac{y}{x} = +\infty$, for $x = 0$ and $y \in [0,1]$.

Example 3. Using the Łukasiewicz t-norm and t-conorm given in Example 1 the following t-norm, t-conorm and implication on $\mathscr{L}^I$ can be constructed. For all x, y in L^I,

$$\mathscr{T}_{T_L}(x,y) = [\max(0,x_1+y_1-1),\max(0,x_1+y_2-1,x_2+y_1-1)],$$
$$\mathscr{S}_{S_L}(x,y) = [\min(1,x_1+y_2,x_2+y_1),\min(1,x_2+y_2)],$$
$$\mathscr{I}_{\mathscr{T}_{T_L}}(x,y) = \mathscr{I}_{\mathscr{S}_{S_L},\mathscr{N}_s}(x,y) = [\min(1,y_1+1-x_1,y_2+1-x_2),\min(1,y_2+1-x_1)],$$

where $\mathscr{N}_s$ is the standard negation on $\mathscr{L}^I$ defined by $\mathscr{N}_s([x_1,x_2]) = [1-x_2,1-x_1]$, for all $[x_1,x_2] \in L^I$.

Example 4. [23] Let T be an arbitrary t-norm on $([0,1],\leq)$ and $t \in [0,1]$. The residual implication $\mathscr{I}_{\mathscr{T}_{T,t}}$ of $\mathscr{T}_{T,t}$ is given by, for all x, y in L^I,

$$\mathscr{I}_{\mathscr{T}_{T,t}}(x,y) = [\min(I_T(x_1,y_1),I_T(x_2,y_2)),\min(I_T(T(x_2,t),y_2),I_T(x_1,y_2))].$$

Proposition 2. [22] *Let $\mathscr{N}$ be a negation on $\mathscr{L}^I$. Then $\mathscr{N}$ is involutive if and only if there exists an involutive negation N on $([0,1],\leq)$ such that, for all $x \in L^I$,*

$$\mathscr{N}(x) = [N(x_2),N(x_1)].$$

Definition 8. For any negation $\mathscr{N}$ on $\mathscr{L}^I$, if there exists negations N_1 and N_2 on $([0,1],\leq)$ with $N_1 \leq N_2$ such that $\mathscr{N}(x) = [N_1(x_2),N_2(x_1)]$, for all $x \in L^I$, then we denote $\mathscr{N}$ by $\mathscr{N}_{N_1,N_2}$, we call $\mathscr{N}$ n-representable and we call N_1 and N_2 the representatives of $\mathscr{N}$.

Note that $\mathscr{N}_{N,N}$ is the best interval representation of N. Furthermore, if N is continuous, then

$$\mathscr{N}_{N,N}([x_1,x_2]) = \{N(x) \mid x \in [x_1,x_2]\}.$$

Proposition 3. [19, 22] *A mapping $\Phi : L^I \to L^I$ is an increasing permutation of $\mathscr{L}^I$ with increasing inverse if and only if there exists an increasing permutation ϕ of $([0,1],\leq)$ such that, for all $x \in L^I$,*

$$\Phi(x) = [\phi(x_1),\phi(x_2)].$$

Let $n \in \mathbb{N} \setminus \{0\}$. If for an n-ary mapping f on $[0,1]$ and an n-ary mapping F on L^I it holds that $F([a_1,a_1],\ldots,[a_n,a_n]) = [f(a_1,\ldots,a_n),f(a_1,\ldots,a_n)]$, for all $(a_1,\ldots,a_n) \in [0,1]^n$, then we say that F is a natural extension of f to L^I. Clearly, for any mapping F on L^I, $F(D,\ldots,D) \subseteq D$ if and only if there exists a mapping f on $[0,1]$ such that F is a natural extension of f to L^I. E.g. for any t-norm T on $([0,1],\leq)$, the t-norms $\mathscr{T}_{T,T}$ and $\mathscr{T}_T$ are natural extensions of T to L^I; if $\mathscr{N}$ is an involutive negation on $\mathscr{L}^I$, then from Proposition 2 it follows that there exists an involutive negation N on $([0,1],\leq)$ such that $\mathscr{N}$ is a natural extension of N.

2.3 Continuity on $\mathscr{L}^I$

In order to introduce continuity on $\mathscr{L}^I$ we need a metric on L^I. Well-known metrics include the Euclidean distance, the Hamming distance and the Moore distance. In the two-dimensional space $\mathbb{R}^2$ they are defined as follows:

- the Euclidean distance between two points $x = (x_1,x_2)$ and $y = (y_1,y_2)$ in $\mathbb{R}^2$ is given by

$$d^E(x,y) = \sqrt{(x_1 - y_1)^2 + (x_2 - y_2)^2}\,,$$

- the Hamming distance between two points $x = (x_1,x_2)$ and $y = (y_1,y_2)$ in $\mathbb{R}^2$ is given by

$$d^H(x,y) = |x_1 - y_1| + |x_2 - y_2|,$$

- the Moore distance between two points $x = (x_1,x_2)$ and $y = (y_1,y_2)$ in $\mathbb{R}^2$ is given by [39]

$$d^M(x,y) = \max(|x_1 - y_1|, |x_2 - y_2|).$$

If we restrict these distances to L^I then we obtain the metric spaces (L^I, d^E), (L^I, d^H) and (L^I, d^M). Note that these distances are homeomorphic when used on $\mathbb{R}^2$ (see [14]). Therefore, the relative topologies w.r.t. L^I are also homeomorphic, which implies that they determine the same set of continuous functions. From now on, if we talk about continuity in L^I, then we mean continuity w.r.t. one of these metric spaces.

It is shown in [22] that for t-norms on $\mathscr{L}^I$ the residuation principle is *not* equivalent to the left-continuity and not even to the continuity of the t-norm: all t-norms on $\mathscr{L}^I$ which satisfy the residuation principle are left-continuous, but the converse does not hold.

3 Representation of Implications on $\mathscr{L}^I$

Similarly as for t-norms we can introduce a direct representability for implications on $\mathscr{L}^I$, as well as optimistic and pessimistic representability, by means of implications on $([0,1], \leq)$.

Lemma 2. *[21] Given implications I_1 and I_2 on $([0,1], \leq)$ with $I_1 \leq I_2$, the mapping $\mathscr{I}_{I_1,I_2} : (L^I)^2 \to L^I$ defined by, for all x,y in L^I,*

$$\mathscr{I}_{I_1,I_2}(x,y) = [I_1(x_2,y_1), I_2(x_1,y_2)]$$

is an implication on $\mathscr{L}^I$.

Note that $\mathscr{I}_{I,I}$ is the best interval representation of I. Furthermore, for continuous I it holds that

$$\mathscr{I}_{I,I}([x_1,x_2],[y_1,y_2]) = \{I(\alpha,\beta) \mid \alpha \in [x_1,x_2] \text{ and } \beta \in [y_1,y_2]\}.$$

Lemma 3. *[21] Given an implication I on $([0,1], \leq)$ the mappings $\mathscr{I}_I : (L^I)^2 \to L^I$ and $\mathscr{I}'_I : (L^I)^2 \to L^I$ defined by, for all x,y in L^I,*

$$\mathscr{I}_I = [I(x_2,y_1), \max(I(x_1,y_1), I(x_2,y_2))],$$
$$\mathscr{I}'_I = [\min(I(x_1,y_1), I(x_2,y_2)), I(x_1,y_2)],$$

are implications on $\mathscr{L}^I$.

Definition 9. [21] Let I_1, I_2 and I be implications on $([0,1], \leq)$, the mappings $\mathscr{I}_{I_1,I_2}$, $\mathscr{I}_I$ and $\mathscr{I}_I'$ defined in Lemma 2 and 3 are called the i-representable implication on $\mathscr{L}^I$ with representatives I_1 and I_2, the pessimistic and the optimistic implication on $\mathscr{L}^I$ with representative I, respectively.

Implications on $\mathscr{L}^I$ can also be generated from t-(co)norms and negations as S- and R-implications. We study the relationship of these constructs to i-representability and optimistic and pessimistic representability.

The following proposition shows that there exists a strong relationship between S-implications on $\mathscr{L}^I$ based on a t-representable t-conorm and S-implications on the unit interval based on the representatives of that t-conorm.

Proposition 4. [6] A mapping $\mathscr{I} : (L^I)^2 \to L^I$ is an S-implication based on an involutive negation $\mathscr{N}_{N,N}$ and on a t-representable t-conorm $\mathscr{S}_{S_1,S_2}$ if, and only if, there exist S-implications $I_{S_1,N}, I_{S_2,N} : [0,1]^2 \to [0,1]$ based on the negation N and the t-conorms S_1 and S_2 respectively, such that

$$\mathscr{I}([x_1,x_2],[y_1,y_2]) = [I_{S_1,N}(x_2,y_1), I_{S_2,N}(x_1,y_2)].$$

So, S-implications on $\mathscr{L}^I$ generated by a t-representable t-conorm and an involutive negation are i-representable implications having an S-implication on $([0,1], \leq)$ as their representative. For R-implications, no such transparent relation with i-representability exists.

Proposition 5. [21] No R-implication on $\mathscr{L}^I$ is i-representable.

We discuss now how optimistic and pessimistic implications can be related to optimistic and pessimistic t-norms through the construction of the corresponding R- and S-implications.

Proposition 6. [21] Let $\mathscr{T}_T$ be a pessimistic t-norm on $\mathscr{L}^I$. Then the R-implication generated by $\mathscr{T}_T$ is given by the optimistic implication with representative I_T, i.e.

$$\mathscr{I}_{\mathscr{T}_T} = \mathscr{I}_{I_T}'.$$

Proposition 7. [21] Let $\mathscr{T}_T'$ be an optimistic t-norm on $\mathscr{L}^I$. Then the R-implication generated by $\mathscr{T}_T'$ is given by, for all x, y in L^I,

$$\mathscr{I}_{\mathscr{T}_T'}(x,y) = [\min(I_T(x_1,y_1), I_T(x_2,y_2)), I_T(x_2,y_2)].$$

This formula resembles the one corresponding to optimistic implications. However, the upper bound involves x_2 instead of x_1, so contrary to optimistic implicators this bound does not correspond to the highest possible value of $I(\alpha,\beta)$, where α, β in $[0,1]$. Obviously, $\mathscr{I}_{\mathscr{T}_T'}$ is not a pessimistic implication either. Moreover, it is equal to the R-implication generated by the corresponding t-representable t-norm $\mathscr{T}_{T,T}$. More generally, we have the following property.

Proposition 8. [21] Let $\mathscr{T}_{T_1,T_2}$ be a t-representable t-norm on $\mathscr{L}^I$. Then the R-implication generated by $\mathscr{T}_{T_1,T_2}$ is given by, for all x, y in L^I,

$$\mathscr{I}_{\mathscr{T}_{T_1,T_2}}(x,y) = [\min(I_{T_1}(x_1,y_1), I_{T_2}(x_2,y_2)), I_{T_2}(x_2,y_2)].$$

For the S-implications corresponding to pessimistic and optimistic t-conorms we obtain the following.

Proposition 9. *[21] Let $\mathscr{S}_S$ be a pessimistic t-conorm on $\mathscr{L}^I$ with representative S and let $\mathscr{N}_{N,N}$ be an n-representable negation with representative N. Then the S-implication generated by $\mathscr{S}_S$ and $\mathscr{N}_{N,N}$ is the pessimistic implication with representative $I_{S,N}$, i.e.*

$$\mathscr{I}_{\mathscr{S}_S,\mathscr{N}_{N,N}} = \mathscr{I}_{I_{S,N}}.$$

Let $\mathscr{S}_S'$ be an optimistic t-conorm on $\mathscr{L}^I$ with representative S and let $\mathscr{N}_{N,N}$ be an n-representable negation with representative N. Then the S-implication generated by $\mathscr{S}_S'$ and $\mathscr{N}_{N,N}$ is the optimistic implication with representative $I_{S,N}$, i.e.

$$\mathscr{I}_{\mathscr{S}_S',\mathscr{N}_{N,N}} = \mathscr{I}_{I_{S,N}}'.$$

We see that pessimistic t-norms generate optimistic R-implications, but optimistic t-norms do not generate pessimistic implications. The R-implications generated by optimistic t-norms coincide with the R-implications generated by t-representable t-norms. However, no intuitive interpretation of these R-implications can be given. On the other hand, for S-implications the situation is clearer: pessimistic t-conorms generate pessimistic S-implications, optimistic t-conorms generate optimistic S-implications and t-representable t-conorms generate i-representable S-implications.

4   Smets-Magrez Axioms

In the previous section we have seen that the class of pessimistic t-norms is the only one which generate both R- and S-implications that belong to one of the classes of representable implications which we discussed before. The superiority of the pessimistic t-norms goes even further as we will see below.

Let $\mathscr{I}$ be an implication on $\mathscr{L}$. The mapping $\mathscr{N}_{\mathscr{I}} : L \to L$ defined by $\mathscr{N}_{\mathscr{I}}(x) = \mathscr{I}(x, 0_{\mathscr{L}})$, for all $x \in L$, is a negation on $\mathscr{L}$, called the negation generated by $\mathscr{I}$.

The Smets-Magrez axioms, a set of natural and commonly imposed criteria for implications on the unit interval, can be extended to $\mathscr{L}^I$ as follows [17]. An implication $\mathscr{I}$ on $\mathscr{L}^I$ is said to satisfy the Smets-Magrez axioms if for all x, y, z in L^I,

(A.1) $\mathscr{I}(.,y)$ is decreasing and $\mathscr{I}(x,.)$ is increasing (monotonicity laws),
(A.2) $\mathscr{I}(1_{\mathscr{L}^I}, x) = x$ (neutrality principle),
(A.3) $\mathscr{I}(\mathscr{N}_{\mathscr{I}}(y), \mathscr{N}_{\mathscr{I}}(x)) = \mathscr{I}(x,y)$ (contrapositivity),
(A.4) $\mathscr{I}(x, \mathscr{I}(y,z)) = \mathscr{I}(y, \mathscr{I}(x,z))$ (exchange principle),
(A.5) $\mathscr{I}(x,y) = 1_{\mathscr{L}^I} \iff x \leq_{L^I} y$ (confinement principle),
(A.6) $\mathscr{I}$ is a continuous $(L^I)^2 \to L^I$ mapping (continuity).

Note that according to our definition, any implication on $\mathscr{L}^I$ satisfies (A.1).

Proposition 10. *[17] An S-implication $\mathscr{I}_{\mathscr{S},\mathscr{N}}$ on $\mathscr{L}^I$ satisfies (A.2), (A.3) and (A.4) if and only if $\mathscr{N}$ is involutive.*

Proposition 11. *[17] An S-implication $\mathscr{I}_{\mathscr{S},\mathscr{N}}$ on $\mathscr{L}^I$ satisfies (A.6) as soon as $\mathscr{S}$ and $\mathscr{N}$ are continuous.*

The following proposition shows that only studying i-representable implicators reduces the possibilities of finding an implication on $\mathscr{L}^I$ which satisfies all Smets-Magrez axioms.

Proposition 12. *[17] Axiom (A.5) fails for every S-implication $\mathscr{I}_{\mathscr{S},\mathscr{N}}$ on $\mathscr{L}^I$ for which $\mathscr{S}$ is t-representable and $\mathscr{N}$ is involutive.*

For R-implications on $\mathscr{L}^I$ we have the following results.

Proposition 13. *[17] Every R-implication $\mathscr{I}_{\mathscr{T}}$ on $\mathscr{L}^I$ satisfies (A.2).*

Proposition 14. *[17] An R-implication $\mathscr{I}_{\mathscr{T}}$ on $\mathscr{L}^I$ satisfies (A.5) if and only if there exists for each $x = [x_1, x_2] \in L^I$ a sequence $(\delta_i)_{i \in \mathbb{N} \setminus \{0\}}$ in $\Omega = \{\delta \mid \delta \in L^I$ and $\delta_2 < 1\}$ such that $\lim_{i \to +\infty} \delta_i = 1_{\mathscr{L}^I}$ and*

$$\lim_{i \to +\infty} \mathrm{pr}_1 \, \mathscr{T}(x, \delta_i) = x_1,$$
$$\lim_{i \to +\infty} \mathrm{pr}_2 \, \mathscr{T}(x, \delta_i) = x_2.$$

As a consequence of the last proposition, if $\mathscr{T}$ is a t-norm on $\mathscr{L}^I$ for which $\mathrm{pr}_1 \, \mathscr{T} : (L^I)^2 \to [0,1]$ and $\mathrm{pr}_2 \, \mathscr{T} : (L^I)^2 \to [0,1]$ are left-continuous mappings, then $\mathscr{I}_{\mathscr{T}}$ satisfies (A.5).

Similarly as for S-implications, limiting ourselves to R-implications generated by t-representable t-norms reduces our chances of finding an implication which satisfies all Smets-Magrez axioms.

Proposition 15. *[17] Axiom (A.3) fails for every R-implication $\mathscr{I}_{\mathscr{T}}$ on $\mathscr{L}^I$ for which $\mathscr{T}$ is t-representable.*

Similarly as for t-norms on the unit interval we have the following property.

Proposition 16. *[17] If an implication $\mathscr{I}$ on $\mathscr{L}^I$ satisfies (A.2), (A.3), (A.4), then the mappings $\mathscr{T}_{\mathscr{I}}, \mathscr{S}_{\mathscr{I}} : (L^I)^2 \to L^I$ defined by, for all x, y in L^I,*

$$\mathscr{T}_{\mathscr{I}}(x,y) = \mathscr{N}_{\mathscr{I}}(\mathscr{I}(x, \mathscr{N}_{\mathscr{I}}(y))),$$
$$\mathscr{S}_{\mathscr{I}}(x,y) = \mathscr{I}(\mathscr{N}_{\mathscr{I}}(x), y),$$

are a t-norm and a t-conorm on L^I, respectively.

As a consequence, all implications on $\mathscr{L}^I$ satisfying (A.2), (A.3) and (A.4) are S-implications.

We check for the class of t-norms $\mathscr{T}_{T,t}$ under which conditions the residual implication $\mathscr{I}_{\mathscr{T}_{T,t}}$ satisfies the Smets-Magrez axioms.

Proposition 17. *[27] Let T be a t-norm on $([0,1],\leq)$ and $t \in [0,1]$. The residual implication $\mathscr{I}_{\mathscr{T}_{T,t}}$ of $\mathscr{T}_{T,t}$ satisfies*

- *(A.1) and (A.2);*
- *(A.3) if and only if $t = 1$ and I_T satisfies (A.3);*
- *(A.4) if and only if I_T satisfies (A.4);*
- *(A.5) if and only if I_T satisfies (A.5);*
- *(A.6) as soon as T is continuous and I_T satisfies (A.6).*

The main result of this section says that the implications on $\mathscr{L}^I$ which satisfy all Smets-Magrez axioms and the additional border condition $\mathscr{I}(D,D) \subseteq D$ (which means that all exact intervals are mapped on exact intervals, or, in other words, that an implication can not add uncertainty when there is no uncertainty in the original values) can be fully characterized in terms of the residual implication of the pessimistic extension of the Łukasiewicz t-norm.

Proposition 18. *[17] An implication $\mathscr{I}$ on $\mathscr{L}^I$ satisfies all Smets-Magrez axioms and $\mathscr{I}(D,D) \subseteq D$ if and only if there exists a continuous increasing permutation Φ of L^I with increasing inverse such that for all x,y in L^I,*

$$\mathscr{I}(x,y) = \Phi^{-1}(\mathscr{I}_{\mathscr{T}_{T_L}}(\Phi(x),\Phi(y))).$$

5 Distributivity of Implication Functions over Triangular Norms and Conorms

In this section we discuss the solutions of equations of the following kind:

$$\mathscr{I}(x,g(y,z)) = g(\mathscr{I}(x,y),\mathscr{I}(x,z)),$$

where $\mathscr{I}$ is an implication function on $\mathscr{L}^I$ and g is a t-norm or a t-conorm on $\mathscr{L}^I$. Distributivity of implications on the unit interval over different fuzzy logic connectives has been studied in the recent past by many authors (see [5, 10, 36, 41, 42, 46]). This interest, perhaps, was kickstarted by Combs and Andrews [16] which exploit the classical tautology

$$(p \wedge q) \to r \equiv (p \to r) \vee (q \to r)$$

in their inference mechanism towards reduction in the complexity of fuzzy "If–Then" rules.

We say that a t-norm T on $([0,1],\leq)$ is strict, if it is continuous and strictly monotone, i.e. $T(x,y) < T(x,z)$ whenever $x > 0$ and $y < z$.

Proposition 19. *A function $T : [0,1]^2 \to [0,1]$ is a strict t-norm if and only if there exists a continuous, strictly decreasing function $t : [0,1] \to [0,+\infty]$ with $t(1) = 0$ and $t(0) = +\infty$, which is uniquely determined up to a positive multiplicative constant, such that*

$$T(x,y) = t^{-1}(t(x)+t(y)), \quad \text{for all } (x,y) \in [0,1]^2.$$

The function t is called an additive generator of T.

In order to be able to find the implications on $\mathscr{L}^I$ which are distributive w.r.t. a t-representable t-norm generated from strict t-norms, we consider the following lemma.

Lemma 4. *[8] Let $L^\infty = \{(u_1, u_2) \mid (u_1, u_2) \in [0, +\infty]^2 \text{ and } u_1 \geq u_2\}$. For a function $f : L^\infty \to [0, +\infty]$ the following statements are equivalent:*

1. f satisfies the functional equation

$$f(u_1 + v_1, u_2 + v_2) = f(u_1, u_2) + f(v_1, v_2), \quad \text{for all } (u_1, u_2), (v_1, v_2) \text{ in } L^\infty;$$

2. either $f = 0$, or $f = +\infty$, or

$$f(u_1, u_2) = \begin{cases} 0, & \text{if } u_2 = 0, \\ +\infty, & \text{else,} \end{cases}$$

or

$$f(u_1, u_2) = \begin{cases} 0, & \text{if } u_2 < +\infty, \\ +\infty, & \text{else,} \end{cases}$$

or

$$f(u_1, u_2) = \begin{cases} 0, & \text{if } u_1 = 0, \\ +\infty, & \text{else,} \end{cases}$$

or

$$f(u_1, u_2) = \begin{cases} 0, & \text{if } u_1 = u_2 < +\infty, \\ +\infty, & \text{else,} \end{cases}$$

or

$$f(u_1, u_2) = \begin{cases} 0, & \text{if } u_2 = 0 \text{ and } u_1 < +\infty, \\ +\infty, & \text{else,} \end{cases}$$

or

$$f(u_1, u_2) = \begin{cases} 0, & \text{if } u_1 < +\infty, \\ +\infty, & \text{else,} \end{cases}$$

or there exists a unique $c \in]0, +\infty[$ such that $f(u_1, u_2) = cu_1$, or $f(u_1, u_2) = cu_2$, or

$$f(u_1, u_2) = \begin{cases} cu_1, & \text{if } u_1 = u_2, \\ +\infty, & \text{else,} \end{cases}$$

or

$$f(u_1, u_2) = \begin{cases} cu_2, & \text{if } u_1 < +\infty, \\ +\infty, & \text{else,} \end{cases}$$

or

$$f(u_1, u_2) = \begin{cases} cu_1, & \text{if } u_2 = 0, \\ +\infty, & \text{else,} \end{cases}$$

or

$$f(u_1, u_2) = \begin{cases} c(u_1 - u_2), & \textit{if } u_2 < +\infty, \\ +\infty, & \textit{else,} \end{cases}$$

or there exist unique c_1, c_2 in $]0, +\infty[$ with $c_1 \neq c_2$ such that

$$f(u_1, u_2) = \begin{cases} c_1(u_1 - u_2) + c_2 u_2, & \textit{if } u_2 < +\infty, \\ +\infty, & \textit{else,} \end{cases}$$

for all $(u_1, u_2) \in L^\infty$.

The following proposition detailing the solutions of the distributivity equation follows immediately from the results in [8].

Proposition 20. *Let $\mathscr{T}$ be a t-representable t-norm generated from strict t-norms with generator t_1 and t_2 respectively. If a function $\mathscr{I} : (L^I)^2 \to L^I$ satisfies the equation*

$$\mathscr{I}(x, \mathscr{T}(y, z)) = \mathscr{T}(\mathscr{I}(x, y), \mathscr{I}(x, z)), \quad \textit{for all } (x, y, z) \in (L^I)^3, \qquad (5)$$

then for each $[x_1, x_2] \in L^I$, each of the functions defined by, for all $(a, b) \in L^\infty$,

$$f_{[x_1, x_2]}(a, b) = t_1 \circ \mathrm{pr}_1 \circ \mathscr{I}([x_1, x_2], [t_1^{-1}(a), t_1^{-1}(b)]),$$
$$f^{[x_1, x_2]}(a, b) = t_2 \circ \mathrm{pr}_2 \circ \mathscr{I}([x_1, x_2], [t_2^{-1}(a), t_2^{-1}(b)])$$

satisfies one of the representations given in Lemma 4.

In [7] the functions $\mathscr{I} : (L^I)^2 \to L^I$ which are continuous w.r.t. the second argument and which satisfy (5) are listed in the case that $\mathscr{T}$ is the t-representable t-norm generated from the product t-norm T_P on $([0, 1], \leq)$.

Not all possibilities for f in Lemma 4 yield a mapping $\mathscr{I}$ which returns values in $\mathscr{L}^I$; furthermore the mappings $\mathscr{I}$ that do only return values in $\mathscr{L}^I$ are not all implications on $\mathscr{L}^I$ [8]. The following example shows that there is at least one possibility for f which produces an implication on $\mathscr{L}^I$.

Example 5. Let $[x_1, x_2]$ be arbitrary in L^I. Define for all $(a, b) \in L^\infty$,

$$f_{[x_1, x_2]}(a, b) = x_2 a,$$
$$f^{[x_1, x_2]}(a, b) = x_1 b.$$

We find

$$\mathscr{I}(x, y) = \begin{cases} 1_{\mathscr{L}^I}, & \text{if } x_2 = 0, \\ [t_1^{-1}(x_2 t_1(y_1)), 1], & \text{if } x_1 = 0 < x_2, \\ [t_1^{-1}(x_2 t_1(y_1)), t_2^{-1}(x_1 t_2(y_2))], & \text{otherwise.} \end{cases}$$

Note that $\mathscr{I} = \mathscr{I}_{I_1,I_2}$ where for $i \in \{1,2\}$, I_i is the implication on $([0,1],\leq)$ defined by, for all x,y in $[0,1]$,

$$I_i(x,y) = \begin{cases} 1, & \text{if } x = 0, \\ t_i^{-1}(xt_i(y)), & \text{otherwise.} \end{cases}$$

It can be straightforwardly verified that the implication $\mathscr{I} = \mathscr{I}_{I_1,I_2}$ and the t-representable t-norm generated by strict t-norms with generators t_1 and t_2 satisfy (5). For $T_1 = T_2 = T_P$ we have that $t_1(x) = t_2(x) = -\ln(x)$ (so $t_1^{-1}(a) = t_2^{-1}(a) = e^{-a}$) and we obtain

$$\mathscr{I}(x,y) = \begin{cases} 1_{\mathscr{L}^I}, & \text{if } x_2 = y_1 = 0, \\ [y_1^{x_2},1], & \text{if } x_1 = y_2 = 0 < x_2, \\ [y_1^{x_2},y_2^{x_1}], & \text{otherwise,} \end{cases}$$

This implication resembles the function $\mathscr{I}$ in Example 11 of [7]; however the latter is not increasing in its second argument and therefore not an implication on $\mathscr{L}^I$. Indeed, $\mathscr{I}([0,1],[0,0]) = 1_{\mathscr{L}^I}$ and $\mathscr{I}([0,1],[y_1,y_1]) = [y_1,1] \not\geq_{L^I} \mathscr{I}([0,1],[0,0])$ for any $y_1 \in \,]0,1[$.

6 Interval-Valued Residuated Lattices, Triangle Algebras and Interval-Valued Monoidal Logic

In this section we relate implications on $\mathscr{L}^I$ to a generalization of fuzzy logic to the interval-valued fuzzy case. We first discuss triangle algebras which are special cases of residuated lattices designed for being used in interval-valued fuzzy set theory.

6.1 Interval-Valued Residuated Lattices and Triangle Algebras

We consider special cases of residuated lattices in which new operators are added so that the resulting structure captures the triangular shape of $\mathscr{L}^I$ (and its generalizations). First we recall the definition of a residuated lattice.

Definition 10. [28] A residuated lattice is a structure $\mathscr{L} = (L,\sqcap,\sqcup,*,\Rightarrow,0,1)$ in which $\sqcap$, $\sqcup$, $*$ and $\Rightarrow$ are binary operators on the set L and

- $(L,\sqcap,\sqcup,0,1)$ is a bounded lattice (with 0 as smallest and 1 as greatest element),
- $*$ is commutative and associative, with 1 as neutral element, and
- $x * y \leq z$ iff $x \leq y \Rightarrow z$ for all x, y and z in L (residuation principle).

The binary operations $*$ and $\Rightarrow$ are called product and implication, respectively. We will use the notations $\neg x$ for $x \Rightarrow 0$ (negation), $x \Longleftrightarrow y$ for $(x \Rightarrow y) \sqcap (y \Rightarrow x)$ and x^n for $\underbrace{x * x * \cdots * x}_{n \text{ times}}$.

Definition 11. [49] Given a lattice $\mathscr{L} = (L, \sqcap, \sqcup)$ (called the base lattice), its triangularization $\mathbb{T}(\mathscr{L})$ is the structure $\mathbb{T}(\mathscr{L}) = (\text{Int}(\mathscr{L}), \sqcap, \sqcup)$ defined by

- $\text{Int}(\mathscr{L}) = \{[x_1, x_2] \mid (x_1, x_2) \in L^2 \text{ and } x_1 \le x_2\}$,
- $[x_1, x_2] \sqcap [y_1, y_2] = [x_1 \sqcap y_1, x_2 \sqcap y_2]$,
- $[x_1, x_2] \sqcup [y_1, y_2] = [x_1 \sqcup y_1, x_2 \sqcup y_2]$.

The set $D_{\mathscr{L}} = \{[x, x] \mid x \in L\}$ is called the set of exact elements of $\mathbb{T}(\mathscr{L})$.

Definition 12. [49] An interval-valued residuated lattice (IVRL) is a residuated lattice $(\text{Int}(\mathscr{L}), \sqcap, \sqcup, \odot, \Rightarrow_{\odot}, [0,0], [1,1])$ on the triangularization $\mathbb{T}(\mathscr{L})$ of a bounded lattice $\mathscr{L} = (L, \cap, \cup)$ in which $D_{\mathscr{L}}$ is closed under $\odot$ and $\Rightarrow_{\odot}$, i.e. $[x, x] \odot [y, y] \in D_{\mathscr{L}}$ and $[x, x] \Rightarrow_{\odot} [y, y] \in D_{\mathscr{L}}$ for all x, y in L.

When we add $[0, 1]$ as a constant, and p_v and p_h (defined by $p_v([x_1, x_2]) = [x_1, x_1]$ and $p_h([x_1, x_2]) = [x_2, x_2]$ for all $[x_1, x_2]$ in $\text{Int}(\mathscr{L})$) as unary operators, the structure $(\text{Int}(\mathscr{L}), \sqcap, \sqcup, \odot, \Rightarrow_{\odot}, p_v, p_h, [0, 0], [0, 1], [1, 1])$ is called an extended IVRL.

Example 6. Let T be a left-continuous t-norm on $([0, 1], \min, \max)$, $t \in [0, 1]$. Then $(L^I, \inf, \sup, \mathscr{T}_{T,t}, \mathscr{I}_{\mathscr{T}_{T,t}}, [0, 0], [1, 1])$ is an IVRL.

The triangular norms $\mathscr{T}$ on $\mathscr{L}^I$ satisfying the residuation principle and which satisfy the property that D is closed under $\mathscr{T}$ and $\mathscr{I}_{\mathscr{T}}$ are completely characterized in terms of a t-norm T on the unit interval.

Proposition 21. *[49] Let* $(\text{Int}(\mathscr{L}), \sqcap, \sqcup, \odot, \Rightarrow_{\odot}, [0, 0], [1, 1])$ *be an IVRL and let* $t \in L$, $* : L^2 \to L$ *and* $\Rightarrow : L^2 \to L$ *be defined by*

$$t = \text{pr}_2([0, 1] \odot [0, 1]),$$
$$x * y = \text{pr}_1([x, x] \odot [y, y]),$$
$$x \Rightarrow y = \text{pr}_1([x, x] \Rightarrow_{\odot} [y, y]),$$

for all x, y *in* L. *Then for all* x, y *in* $\text{Int}(\mathscr{L})$,

$$[x_1, x_2] \odot [y_1, y_2] = [x_1 * y_1, (x_2 * y_2 * t) \cup (x_1 * y_2) \cup (x_2 * y_1)],$$
$$[x_1, x_2] \Rightarrow_{\odot} [y_1, y_2] = [(x_1 \Rightarrow y_1) \cap (x_2 \Rightarrow y_2), (x_1 \Rightarrow y_2) \cap (x_2 \Rightarrow (t \Rightarrow y_2))].$$

To capture the triangular structure of IVRLs, we extend the definition of a residuated lattice with a new constant u ("uncertainty") and two new unary connectives v ("necessity") and μ ("possibility"). Intuitively, the elements of a triangle algebra may be thought of as closed intervals, u as the interval $[0, 1]$, and v and μ as operators mapping $[x_1, x_2]$ to $[x_1, x_1]$ and $[x_2, x_2]$ respectively.

Definition 13. [47, 50] A triangular algebra is a structure $\mathscr{A} = (A, \sqcap, \sqcup, *, \Rightarrow, v, \mu, 0, u, 1)$ of type $(2, 2, 2, 2, 1, 1, 0, 0, 0)$ such that $(A, \sqcap, \sqcup, *, \Rightarrow, 0, 1)$ is a residuated lattice and, for all x, y in A,

(T.1) $vx \leq x$,

(T.2) $vx \leq vvx$,

(T.3) $v(x \sqcap y) = vx \sqcap vy$,

(T.4) $v(x \sqcup y) = vx \sqcup vy$,

(T.5) $vu = 0$,

(T.6) $v\mu x = \mu x$,

(T.7) $v(x \Rightarrow y) \leq vx \Rightarrow vy$,

(T.8) $(vx \Leftrightarrow vy) * (\mu x \Leftrightarrow \mu y) \leq (x \Leftrightarrow y)$,

(T.9) $vx \Rightarrow vy \leq v(vx \Rightarrow vy)$,

(T.1') $x \leq \mu x$,

(T.2') $\mu\mu x \leq \mu x$,

(T.3') $\mu(x \sqcap y) = \mu x \sqcap \mu y$,

(T.4') $\mu(x \sqcup y) = \mu x \sqcup \mu y$,

(T.5') $\mu u = 1$,

(T.6') $\mu vx = vx$,

where the biresiduum $\Leftrightarrow$ is defined as $x \Leftrightarrow y = x \Rightarrow y \wedge y \Rightarrow x$, for all x, y in A. The unary operators v and μ are called the necessity and possibility operator, respectively.

Note that in a triangle algebra $x = vx \sqcup (\mu x \sqcap u)$, for all $x \in A$. This shows that an element of the triangle algebra is completely determined by its necessity and its possibility.

There is a one-to-one correspondance between the class of IVRLs and the class of triangle algebras.

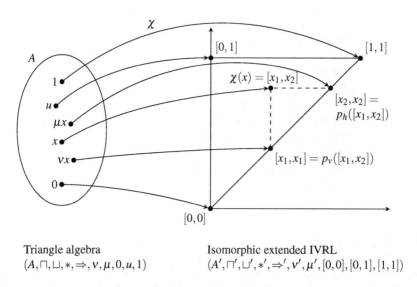

Triangle algebra
$(A, \sqcap, \sqcup, *, \Rightarrow, v, \mu, 0, u, 1)$

Isomorphic extended IVRL
$(A', \sqcap', \sqcup', *', \Rightarrow', v', \mu', [0,0], [0,1], [1,1])$

Fig. 2 The isomorphism ξ from a triangle algebra to an IVRL

Proposition 22. *[50] Every triangle algebra is isomorphic to an extended IVRL (see Figure 2). Conversely, every extended IVRL is a triangular algebra.*

6.2 Interval-Valued Monoidal Logic

We now translate the defining properties of triangle algebras into logical axioms and show that the resulting logic IVML is sound and complete w.r.t. the variety of triangle algebras.

The language of IVML consists of countably many proposition variables (p_1, p_2, ...), the constants $\overline{0}$ and $\overline{u}$, the unary operators $\square$, $\lozenge$, the binary operators $\wedge$, $\vee$, &, $\rightarrow$, and finally the auxiliary symbols '(' and ')'. Formulas are defined inductively: proposition variables, $\overline{0}$ and $\overline{u}$ are formulas; if ϕ and ψ are formulas, then so are $(\phi \wedge \psi)$, $(\phi \vee \psi)$, $(\phi \& \psi)$, $(\phi \rightarrow \psi)$, $(\square \psi)$ and $(\lozenge \psi)$.

In order to avoid unnecessary brackets, we agree on the following priority rules:

- unary operators always take precedence over binary ones, while
- among the binary operators, & has the highest priority; furthermore $\wedge$ and $\vee$ take precedence over $\rightarrow$,
- the outermost brackets are not written.

We also introduce some useful shorthand notations: $\overline{1}$ for $\overline{0} \rightarrow \overline{0}$, $\neg \phi$ for $\phi \rightarrow \overline{0}$ and $\phi \leftrightarrow \psi$ for $(\phi \rightarrow \psi) \wedge (\psi \rightarrow \phi)$ for formulas ϕ and ψ.

The axioms of IVML are those of ML (Monoidal Logic) [35], i.e.

$$(ML.1) \quad (\phi \rightarrow \psi) \rightarrow ((\psi \rightarrow \chi) \rightarrow (\phi \rightarrow \chi)),$$
$$(ML.2) \quad \phi \rightarrow (\phi \vee \psi),$$
$$(ML.3) \quad \psi \rightarrow (\phi \vee \psi),$$
$$(ML.4) \quad (\phi \rightarrow \chi) \rightarrow ((\psi \rightarrow \chi) \rightarrow ((\phi \vee \psi) \rightarrow \chi)),$$
$$(ML.5) \quad (\phi \wedge \psi) \rightarrow \phi,$$
$$(ML.6) \quad (\phi \wedge \psi) \rightarrow \psi,$$
$$(ML.7) \quad (\phi \& \psi) \rightarrow \phi,$$
$$(ML.8) \quad (\phi \& \psi) \rightarrow (\psi \& \phi),$$
$$(ML.9) \quad (\phi \rightarrow \psi) \rightarrow ((\phi \rightarrow \chi) \rightarrow (\phi \rightarrow (\psi \wedge \chi))),$$
$$(ML.10) \, (\phi \rightarrow (\psi \rightarrow \chi)) \rightarrow ((\phi \& \psi) \rightarrow \chi),$$
$$(ML.11) \, ((\phi \& \psi) \rightarrow \chi) \rightarrow (\phi \rightarrow (\psi \rightarrow \chi)),$$
$$(ML.12) \, \overline{0} \rightarrow \phi,$$

complemented with

$$(IVML.1) \quad \square \phi \rightarrow \phi, \qquad\qquad (IVML.1') \quad \phi \rightarrow \lozenge \phi,$$
$$(IVML.2) \quad \square \phi \rightarrow \square \square \phi, \qquad\qquad (IVML.2') \quad \lozenge \lozenge \phi \rightarrow \lozenge \phi,$$
$$(IVML.3) \quad (\square \phi \wedge \square \psi) \rightarrow \square(\phi \wedge \psi), \qquad (IVML.3') \quad (\lozenge \phi \wedge \lozenge \psi) \rightarrow \lozenge(\phi \wedge \psi),$$
$$(IVML.4) \quad \square(\phi \vee \psi) \rightarrow (\square \phi \vee \square \psi), \qquad (IVML.4') \quad \lozenge(\phi \vee \psi) \rightarrow (\lozenge \phi \vee \lozenge \psi),$$
$$(IVML.5) \quad \square \overline{1}, \qquad\qquad (IVML.5') \quad \neg \lozenge \overline{0},$$
$$(IVML.6) \quad \neg \square \overline{u}, \qquad\qquad (IVML.6') \quad \lozenge \overline{u},$$
$$(IVML.7) \quad \lozenge \phi \rightarrow \square \lozenge \phi, \qquad\qquad (IVML.7') \quad \lozenge \square \phi \rightarrow \square \phi,$$
$$(IVML.8) \quad \square(\phi \rightarrow \psi) \rightarrow (\square \phi \rightarrow \square \psi),$$
$$(IVML.9) \quad (\square \phi \leftrightarrow \square \psi) \& (\lozenge \phi \leftrightarrow \lozenge \psi) \rightarrow (\phi \leftrightarrow \psi),$$
$$(IVML.10) \, (\square \phi \rightarrow \square \psi) \rightarrow \square(\square \phi \rightarrow \square \psi).$$

The deduction rules are modus ponens (MP, from ϕ and $\phi \to \psi$ infer ψ), generalization[2] (G, from ϕ infer $\Box\phi$) and monotonicity of $\Diamond$ (M$\Diamond$, from $\phi \to \psi$ infer $\Diamond\phi \to \Diamond\psi$).

The consequence relation $\vdash$ is defined as follows, in the usual way. Let V be a theory, i.e., a set of formulas in IVML. A (formal) proof of a formula ϕ in V is a finite sequence of formulas with ϕ at its end, such that every formula in the sequence is either an axiom of IVML, a formula of V, or the result of an application of an inference rule to previous formulas in the sequence. If a proof for ϕ exists in V, we say that ϕ can be deduced from V and we denote this by $V \vdash \phi$.

For a theory V, and formulas ϕ and ψ in IVML, denote $\phi \sim_V \psi$ iff $V \vdash \phi \to \psi$ and $V \vdash \psi \to \phi$ (this is also equivalent with $V \vdash \phi \leftrightarrow \psi$).

Note that (IVML.5) is in fact superfluous, as it immediately follows from $\emptyset \vdash \overline{1}$ and generalization; we include it here to obtain full correspondence with Definition 13.

Definition 14. Let $\mathscr{A} = (A, \sqcap, \sqcup, *, \Rightarrow, v, \mu, 0, u, 1)$ be a triangle algebra and V a theory. An $\mathscr{A}$-evaluation is a mapping e from the set of formulas of IVML to A that satisfies, for each two formulas ϕ and ψ: $e(\phi \wedge \psi) = e(\phi) \sqcap e(\psi)$, $e(\phi \vee \psi) = e(\phi) \sqcup e(\psi)$, $e(\phi \& \psi) = e(\phi) * e(\psi)$, $e(\phi \to \psi) = e(\phi) \Rightarrow e(\psi)$, $e(\Box\phi) = ve(\phi)$, $e(\Diamond\phi) = \mu e(\phi)$, $e(\overline{0}) = 0$ and $e(\overline{u}) = u$. If an $\mathscr{A}$-evaluation e satisfies $e(\chi) = 1$ for every χ in V, it is called an $\mathscr{A}$-model for V.

The following property shows that interval-valued monoidal logic is sound w.r.t. the variety of triangle algebras, i.e., that if a formula ϕ can be deduced from a theory V in IVML, then for every triangle algebra $\mathscr{A}$ and for every $\mathscr{A}$-model e of V, $e(\phi) = 1$, and that IVML is also complete (i.e. that the converse of soundness also holds).

Proposition 23 (Soundness and completeness of IVML). *[50] A formula ϕ can be deduced from a theory V in IVML iff for every triangle algebra $\mathscr{A}$ and for every $\mathscr{A}$-model e of V, $e(\phi) = 1$.*

By adding axioms, we can obtain axiomatic extensions of interval-valued monoidal logic. For these extensions a similar soundness and completeness result holds. Furthermore, in some cases we can obtain a stronger result. For example, similarly as for monoidal t-norm based logic (MTL) [32] we obtain the following result.

Proposition 24 (Standard completeness). *[51] For each formula ϕ, the following three statements are equivalent:*

- *ϕ can be deduced from a theory V in IVMTL (which is the axiomatic extension of IVML obtained by adding the axiom scheme $(\Box\phi \to \Box\psi) \vee (\Box\psi \to \Box\phi)$),*
- *for every triangle algebra $\mathscr{A}$ in which the set of exact elements is prelinear and for every $\mathscr{A}$-model e of V, $e(\phi) = 1$,*
- *for every triangle algebra $\mathscr{A}$ in which the set of exact elements is linear and for every $\mathscr{A}$-model e of V, $e(\phi) = 1$.*

---

[2] Generalization is often called necessitation, e.g. in [59].

More information on this and on the soundness and completeness of other axiomatic extensions of IVML can be found in [47, 51, 52].

Remark 1. Interval-valued monoidal logic is a truth-functional logic: the truth degree of a compound proposition is determined by the truth degree of its parts. This causes some counterintuitive results, if we want to interpret the element $[0,1]$ of an IVRL as uncertainty. For example: suppose we don't know anything about the truth value of propositions p and q, i.e., $v(p) = v(q) = [0,1]$. Then yet the implication $p \to q$ is definitely valid: $v(p \to q) = v(p) \Rightarrow v(q) = [1,1]$. However, if $\neg[0,1] = [0,1]$ [3] (which is intuitively preferable, since the negation of an uncertain proposition is still uncertain), then we can take $q = \neg p$, and obtain that $p \to \neg p$ is true. Or, equivalently (using the residuation principle), that $p \& p$ is false. This does not seem intuitive, as one would rather expect $p \& p$ to be uncertain if p is uncertain. Another consequence of $[0,1] \Rightarrow [0,1] = [1,1]$ is that it is impossible to interpret the intervals as a set in which the 'real' (unknown) truth value is contained, and $X \Rightarrow Y$ as the smallest closed interval containing every $x \Rightarrow y$, with x in X and y in Y (as in [31]). Indeed: $1 \in [0,1]$ and $0 \in [0,1]$, but $1 \Rightarrow 0 = 0 \notin [1,1]$.

On the other hand, for t-norms it is possible that $X * Y$ is the smallest closed interval containing every $x * y$, with x in X and y in Y, but only if they are t-representable (described by the axiom $\mu(x * y) = \mu x * \mu y$). However, in this case $\neg[0,1] = [0,0]$, which does not seem intuitive ('the negation of an uncertain proposition is absolutely false').

These considerations seem to suggest that interval-valued monoidal logic is not suitable to reason with uncertainty. This does not mean that intervals are not a good way for representing degrees of uncertainty, only that they are not suitable as truth values in a truth functional logical calculus when we interpret them as expressing uncertainty. It might even be impossible to model uncertainty as a truth value in any truth-functional logic. This question is discussed in [29, 30]. However, nothing prevents the intervals in interval-valued monoidal logic from having more adequate interpretations.

7   Other Constructions of Implications on $\mathscr{L}^I$

In the previous section we have seen that implications on $\mathscr{L}^I$ can be constructed as R- or S-implications derived from t-norms or t-conorms on $\mathscr{L}^I$. In this section we describe several other constructions of implications on $\mathscr{L}^I$.

7.1   Conjugacy

In Proposition 3 we have seen that an increasing permutation on $\mathscr{L}^I$ which has an increasing inverse is completely determined by a permutation on $([0,1], \leq)$. Such permutations can be used to construct new functions as follows.

---

[3] This is for example the case if $\neg$ is involutive.

We say that the functions $F, G : (L^I)^2 \to L^I$ are conjugate, if there exists an increasing bijection $\Phi : L^I \to L^I$ with increasing inverse such that $G = F_\Phi$, where

$$F_\phi(x,y) = \Phi^{-1}(F(\Phi(x), \Phi(y))), \quad \text{for all } (x,y) \in (L^I)^2.$$

Proposition 25. *[6] Let $\Phi : L^I \to L^I$ be an increasing permutation with increasing inverse.*

- *If $\mathscr{I}$ is an implication on $\mathscr{L}^I$, then $\mathscr{I}_\Phi$ is an implication on $\mathscr{L}^I$.*
- *If $\mathscr{I}$ is an S-implication on $\mathscr{L}^I$ based on some t-conorm $\mathscr{S}$ and strong negation $\mathscr{N}$ on $\mathscr{L}^I$, then $\mathscr{I}_\Phi$ is also an S-implication on $\mathscr{L}^I$ based on the t-conorm $\mathscr{S}_\Phi$ and strong negation $\mathscr{N}_\Phi$.*
- *If $\mathscr{I}$ is an R-implication on $\mathscr{L}^I$ based on some t-norm $\mathscr{T}$ on $\mathscr{L}^I$, then $\mathscr{I}_\Phi$ is also an R-implication on $\mathscr{L}^I$ based on the t-norm $\mathscr{T}_\Phi$.*

7.2 Implications Defined Using Arithmetic Operators on $\mathscr{L}^I$

Let $\bar{L}^I = \{[x_1, x_2] \mid (x_1, x_2) \in \mathbb{R}^2 \text{ and } x_1 \leq x_2\}$ and $\bar{L}^I_+ = \{[x_1, x_2] \mid (x_1, x_2) \in [0, +\infty[^2 \text{ and } x_1 \leq x_2\}$. We start from two arithmetic operators $\oplus : (\bar{L}^I)^2 \to \bar{L}^I$ and $\otimes : (\bar{L}^I_+)^2 \to \bar{L}^I$ satisfying the following properties (see [20]),

(ADD-1) $\oplus$ is commutative,
(ADD-2) $\oplus$ is associative,
(ADD-3) $\oplus$ is increasing,
(ADD-4) $0_{\mathscr{L}^I} \oplus a = a$, for all $a \in \bar{L}^I$,
(ADD-5) $[\alpha, \alpha] \oplus [\beta, \beta] = [\alpha + \beta, \alpha + \beta]$, for all α, β in $\mathbb{R}$,
(MUL-1) $\otimes$ is commutative,
(MUL-2) $\otimes$ is associative,
(MUL-3) $\otimes$ is increasing,
(MUL-4) $1_{\mathscr{L}^I} \otimes a = a$, for all $a \in \bar{L}^I_+$,
(MUL-5) $[\alpha, \alpha] \otimes [\beta, \beta] = [\alpha\beta, \alpha\beta]$, for all α, β in $[0, +\infty[$.

The conditions (ADD-1)–(ADD-4) and (MUL-1)–(MUL-4) are natural conditions for any addition and multiplication operators. The conditions (ADD-5) and (MUL-5) ensure that these operators are natural extensions of the addition and multiplication of real numbers to $\bar{L}^I$.

The mapping $\ominus$ is defined in [20] by, for all x, y in $\bar{L}^I$,

$$1_{\mathscr{L}^I} \ominus x = [1 - x_2, 1 - x_1], \tag{6}$$

$$x \ominus y = 1_{\mathscr{L}^I} \ominus ((1_{\mathscr{L}^I} \ominus x) \oplus y). \tag{7}$$

Similarly, the mapping $\oslash$ is defined by, for all x, y in $\bar{L}^I_{+,0}$,

$$1_{\mathscr{L}^I} \oslash x = \left[\frac{1}{x_2}, \frac{1}{x_1}\right], \tag{8}$$

$$x \oslash y = 1_{\mathscr{L}^I} \oslash ((1_{\mathscr{L}^I} \oslash x) \otimes y). \tag{9}$$

The properties of these operators are studied in [20].

Using the arithmetic operators on $\mathscr{L}^I$ we can construct t-norms, t-conorms and implications on $\mathscr{L}^I$ which are generalizations of the Łukasiewicz t-norm, t-conorm and implication on the unit interval and which have a similar arithmetic expression as those functions.

Proposition 26. *[20] The mapping $\mathscr{S}_\oplus : (L^I)^2 \to L^I$ defined by, for all x,y in L^I,*

$$\mathscr{S}_\oplus(x,y) = \inf(1_{\mathscr{L}^I}, x \oplus y), \tag{10}$$

is a t-conorm on $\mathscr{L}^I$ if and only if $\oplus$ satisfies the following condition:

$$(\forall(x,y,z) \in (L^I)^3)$$
$$\left(\left((\inf(1_{\mathscr{L}^I}, x \oplus y) \oplus z)_1 < 1 \text{ and } (x \oplus y)_2 > 1\right)\right. \tag{11}$$
$$\left. \Longrightarrow (\inf(1_{\mathscr{L}^I}, x \oplus y) \oplus z)_1 = (x \oplus \inf(1_{\mathscr{L}^I}, y \oplus z))_1\right).$$

Furthermore $\mathscr{S}_\oplus$ is a natural extension of S_L to L^I.

Proposition 27. *[20] The mapping $\mathscr{T}_\oplus : (L^I)^2 \to L^I$ defined by, for all x,y in L^I,*

$$\mathscr{T}_\oplus(x,y) = \sup(0_{\mathscr{L}^I}, x \ominus (1_{\mathscr{L}^I} \ominus y)), \tag{12}$$

is a t-norm on $\mathscr{L}^I$ if and only if $\oplus$ satisfies (11). Furthermore, $\mathscr{T}_\oplus$ is a natural extension of T_L to L^I.

The following theorem gives a simpler sufficient condition so that $\mathscr{S}_\oplus$ is a t-conorm and $\mathscr{T}_\oplus$ is a t-norm on $\mathscr{L}^I$.

Proposition 28. *[20] Assume that $\oplus$ satisfies the following condition:*

$$(\forall(x,y) \in \bar{L}^I_+ \times L^I)$$
$$\left(\left(([x_1,1] \oplus y)_1 < 1 \text{ and } x_2 \in]1,2]\right) \Longrightarrow ([x_1,1] \oplus y)_1 = (x \oplus y)_1\right). \tag{13}$$

Then the mappings $\mathscr{T}_\oplus, \mathscr{S}_\oplus : (L^I)^2 \to L^I$ defined by, for all x,y in L^I,

$$\mathscr{T}_\oplus(x,y) = \sup(0_{\mathscr{L}^I}, x \ominus (1_{\mathscr{L}^I} \ominus y)),$$
$$\mathscr{S}_\oplus(x,y) = \inf(1_{\mathscr{L}^I}, x \oplus y),$$

are a t-norm and a t-conorm on $\mathscr{L}^I$ respectively. Furthermore $\mathscr{T}_\oplus$ is a natural extension of T_L to L^I, and $\mathscr{S}_\oplus$ is a natural extension of S_L to L^I.

Note that $\mathscr{N}_s(x) = 1_{\mathscr{L}^I} \ominus x$, for all $x \in L^I$. So we obtain the following result.

Proposition 29. *Under the same conditions as in Proposition 26 or 28, the mapping $\mathscr{I}_{\mathscr{S}_\oplus, \mathscr{N}_s}$ defined by, for all x,y in L^I,*

$$\mathscr{I}_{\mathscr{S}_{\oplus},\mathscr{N}_s}(x,y) = \mathscr{S}_{\oplus}(\mathscr{N}_s(x),y) = \inf(1_{\mathscr{L}^I},(1_{\mathscr{L}^I} \ominus x) \oplus y),$$

is an implication on $\mathscr{L}^I$. Furthermore, $\mathscr{I}_{\mathscr{S}_{\oplus},\mathscr{N}_s}$ is a natural extension of $\mathscr{I}_{S_L,N_s}$ to L^I.

7.3 Implications Generated by Binary Aggregation Functions

By modifying the definition of $\mathscr{T}_{T,T}$, $\mathscr{T}_T$, ... we can obtain new binary aggregation functions on $\mathscr{L}^I$ which are not t-norms or t-conorms. For example, let T and T' be t-norms and S and S' be t-conorms on $([0,1],\leq)$ with $T \leq T'$ and $S \leq S'$, then we define for all x,y in L^I (see [26]),

$$\mathscr{A}_T(x,y) = [\min(T(x_1,y_2),T(y_1,x_2)),\max(T(x_1,y_2),T(y_1,x_2))],$$
$$\mathscr{A}_S(x,y) = [\min(S(x_1,y_2),S(y_1,x_2)),\max(S(x_1,y_2),S(y_1,x_2))],$$
$$\mathscr{A}'_{T,T'}(x,y) = [\min(T(x_1,y_2),T(y_1,x_2)),T'(x_2,y_2)],$$
$$\mathscr{A}'_{S,S'}(x,y) = [S(x_1,y_1),\max(S'(x_2,y_1),S'(y_2,x_1))].$$

Proposition 30. *[26, 37] Let T and T' be left-continuous t-norms with $T \leq T'$, S and S' be t-conorms with $S \leq S'$, and N be an involutive negation on $([0,1],\leq)$. Then*

- *the residuum of $\mathscr{A}_T$ is equal to the residual implication of $\mathscr{T}_T$, i.e. $\mathscr{I}_{\mathscr{A}_T} = \mathscr{I}_{\mathscr{T}_T}$;*
- *the mapping $\mathscr{I}_{\mathscr{A}_S,\mathscr{N}} : (L^I)^2 \to L^I$ given by, for all x,y in L^I,*

$$\mathscr{I}_{\mathscr{A}_S,\mathscr{N}_{N,N}}(x,y) = [\min(I_{S,N}(x_1,y_1),I_{S,N}(x_2,y_2)),\max(I_{S,N}(x_1,y_1),I_{S,N}(x_2,y_2))],$$

is an implication on $\mathscr{L}^I$;
- *the residuum of $\mathscr{A}'_{T,T'}$ given by, for all x,y in L^I,*

$$\mathscr{I}_{\mathscr{A}'_{T,S}}(x,y) = [\min(I_T(x_1,y_1),I_{T'}(x_2,y_2)),I_{T'}(x_2,y_2)],$$

is an implication on $\mathscr{L}^I$;
- *the mapping $\mathscr{I}_{\mathscr{A}'_{S,S'},\mathscr{N}_{N,N}} : (L^I)^2 \to L^I$ given by, for all x,y in L^I,*

$$\mathscr{I}_{\mathscr{A}'_{S,S'},\mathscr{N}_{N,N}}(x,y) = [S(N(x_2),y_1),\max(S'(N(x_1),y_1),S'(N(x_2),y_2))],$$

is an implication on $\mathscr{L}^I$.

7.4 Implications Generated by Uninorms on $\mathscr{L}^I$

Similarly as for t-norms, implications can also be derived from uninorms. Uninorms are a generalization of t-norms and t-conorms for which the neutral element can be any element of $\mathscr{L}^I$.

Definition 15. [25] A uninorm on a complete lattice $\mathscr{L} = (L, \leq_L)$ is a commutative, associative, increasing mapping $\mathscr{U} : L^2 \to L$ which satisfies

$$(\exists e \in L)(\forall x \in L)(\mathscr{U}(e,x) = x).$$

The element e corresponding to a uninorm $\mathscr{U}$ is unique and is called the neutral element of $\mathscr{U}$.

If $\mathscr{U}(0_{\mathscr{L}}, 1_{\mathscr{L}}) = 0_{\mathscr{L}}$, then $\mathscr{U}$ is called conjunctive, if $\mathscr{U}(0_{\mathscr{L}}, 1_{\mathscr{L}}) = 1_{\mathscr{L}}$, then $\mathscr{U}$ is called disjunctive. Although all uninorms on the unit interval are either conjunctive or disjunctive [56], this is not the case anymore for uninorms on $\mathscr{L}^I$ [18].

Now we construct R- and S-implications derived from uninorms on $\mathscr{L}^I$.

Proposition 31. *[25] Let $\mathscr{U}$ be a uninorm on $\mathscr{L}^I$ with neutral element $e \in L^I \setminus \{0_{\mathscr{L}^I}, 1_{\mathscr{L}^I}\}$. Let $\Omega = \{\omega \mid \omega \in L^I \text{ and } \omega_2 > 0\}$. The mapping $\mathscr{I}_{\mathscr{U}} : (L^I)^2 \to L^I$ defined by, for all x,y in L^I,*

$$\mathscr{I}_{\mathscr{U}}(x,y) = \sup\{z \mid z \in L^I \text{ and } \mathscr{U}(x,z) \leq_{L^I} y\}$$

is an implication on $\mathscr{L}^I$ if and only if

$$(\forall \omega \in \Omega)(\mathscr{U}(0_{\mathscr{L}^I}, \omega) = 0_{\mathscr{L}^I}).$$

As a consequence of this proposition, if $\mathscr{U}$ is conjunctive, then $\mathscr{I}_{\mathscr{U}}$ is an implication on $\mathscr{L}^I$. Note also that $\mathscr{I}_{\mathscr{U}}(e,x) = x$, for all $x \in L^I$.

Proposition 32. *[25] Let $\mathscr{U}$ be a uninorm and $\mathscr{N}$ a negation on $\mathscr{L}^I$. Then the mapping $\mathscr{I}_{\mathscr{U}, \mathscr{N}} : (L^I)^2 \to L^I$ defined by, for all x,y in L^I,*

$$\mathscr{I}_{\mathscr{U}, \mathscr{N}}(x,y) = \mathscr{U}(\mathscr{N}(x), y)$$

is an implication on $\mathscr{L}^I$ if and only if $\mathscr{U}$ is disjunctive.

8   Conclusion

In this work we have listed some results pertaining to implications in interval-valued fuzzy set theory. We have described several possibilities to represent such implications using implications on the unit interval. We gave a characterization of the implications in interval-valued fuzzy set theory which satisfy the Smets-Magrez axioms. We discussed the solutions of a particular distributivity equation involving strict t-norms. We extended monoidal logic to the interval-valued fuzzy case and we gave a soundness and completeness theorem which is similar to the one existing for monoidal logic. Finally we discussed some other constructions of implications in interval-valued fuzzy set theory.

References

1. Atanassov, K.T.: Intuitionistic fuzzy sets. VII ITKR's Session, Sofia (deposed in Central Sci.-Technical Library of Bulg. Acad. of Sci., 1697/84) (1983) (in Bulgarian)
2. Atanassov, K.T.: Intuitionistic fuzzy sets. Fuzzy Sets and Systems 20(1), 87–96 (1986)
3. Atanassov, K.T.: Intuitionistic fuzzy sets. Physica-Verlag, Heidelberg (1999)
4. Atanassov, K.T.: Remarks on the conjunctions, disjunctions and implications of the intuitionistic fuzzy logic. International Journal of Uncertainty, Fuzziness and Knowledge-Based Systems 9(1), 55–65 (2001)
5. Baczyński, M.: On a class of distributive fuzzy implications. International Journal of Uncertainty, Fuzziness and Knowledge-Based Systems 9, 229–238 (2001)
6. Baczyński, M.: On some properties of intuitionistic fuzzy implications. In: Proceedings of the 3rd Conference of the European Society for Fuzzy Logic and Technology (EUSFLAT 2003), Zittau, Germany, pp. 168–171 (2003)
7. Baczyński, M.: On the distributivity of implication operations over t-representable t-norms generated from strict t-norms in atanassov's intuitionistic fuzzy sets theory. In: Magdalena, L., Ojeda-Aciego, M., Verdegay, J.L. (eds.) Proceedings of the 12th International Conference on Information Processing and Management of Uncertainty in Knowledge-Based Systems (IPMU 2008), pp. 1612–1619 (2008)
8. Baczyński, M.: On the Distributivity of Implication Operations over t-Representable t-Norms Generated from Strict t-Norms in Interval-Valued Fuzzy Sets Theory. In: Hüllermeier, E., Kruse, R., Hoffmann, F. (eds.) IPMU 2010. CCIS, vol. 80, pp. 637–646. Springer, Heidelberg (2010)
9. Baczyński, M., Jayaram, B.: Fuzzy implications. STUDFUZZ, vol. 231. Springer, Berlin (2008)
10. Baczyński, M., Jayaram, B.: On the distributivity of fuzzy implications over nilpotent or strict triangular conorms. IEEE Transactions of Fuzzy Systems 17, 590–603 (2009)
11. Bedregal, B.R.C., Dimuro, G.P., Santiago, R.H.N., Reiser, R.H.S.: On interval fuzzy s-implications. Information Sciences 180, 1373–1389 (2010)
12. Bedregal, B.R.C., Takahashi, A.: The best interval representations of t-norms and automorphisms. Fuzzy Sets and Systems 157(24), 3220–3230 (2006)
13. Beliakov, G., Bustince, H., Goswami, D.P., Mukherjee, U.K., Pal, N.R.: On averaging operators for atanassov's intuitionistic fuzzy sets. Information Sciences 181, 1116–1124 (2011)
14. Binmore, K.G.: The foundations of analysis: A straightforward introduction. Book 2 Topological ideas. Cambridge University Press (1981)
15. Bustince, H., Barrenechea, E., Pagola, M., Fernandez, J.: Interval-valued fuzzy sets constructed from matrices: Application to edge detection. Fuzzy Sets and Systems 160(13), 1819–1840 (2009)
16. Combs, W.E., Andrews, J.E.: Combinatorial rule explosion eliminated by a fuzzy rule configuration. IEEE Transactions on Fuzzy Systems 6, 1–11 (1998)
17. Cornelis, C., Deschrijver, G., Kerre, E.E.: Implication in intuitionistic and interval-valued fuzzy set theory: construction, classification and application. International Journal of Approximate Reasoning 35(1), 55–95 (2004)
18. Deschrijver, G.: Uninorms which are neither conjunctive nor disjunctive in interval-valued fuzzy set theory. Information Sciences (submitted)
19. Deschrijver, G.: A thorough study of the basic operators in intuitionistic fuzzy set theory. Ph. D. Thesis, Ghent University, Belgium (2004) (in Dutch)
20. Deschrijver, G.: Generalized arithmetic operators and their relationship to t-norms in interval-valued fuzzy set theory. Fuzzy Sets and Systems 160(21), 3080–3102 (2009)

21. Deschrijver, G., Cornelis, C.: Representability in interval-valued fuzzy set theory. International Journal of Uncertainty, Fuzziness and Knowledge-Based Systems 15(3), 345–361 (2007)
22. Deschrijver, G., Cornelis, C., Kerre, E.E.: On the representation of intuitionistic fuzzy t-norms and t-conorms. IEEE Transactions on Fuzzy Systems 12(1), 45–61 (2004)
23. Deschrijver, G., Kerre, E.E.: Classes of intuitionistic fuzzy t-norms satisfying the residuation principle. International Journal of Uncertainty, Fuzziness and Knowledge-Based Systems 11(6), 691–709 (2003)
24. Deschrijver, G., Kerre, E.E.: On the relationship between some extensions of fuzzy set theory. Fuzzy Sets and Systems 133(2), 227–235 (2003)
25. Deschrijver, G., Kerre, E.E.: Uninorms in L^*-fuzzy set theory. Fuzzy Sets and Systems 148(2), 243–262 (2004)
26. Deschrijver, G., Kerre, E.E.: Implicators based on binary aggregation operators in interval-valued fuzzy set theory. Fuzzy Sets and Systems 153(2), 229–248 (2005)
27. Deschrijver, G., Kerre, E.E.: Smets-Magrez axioms for intuitionistic fuzzy R-implicators. International Journal of Uncertainty, Fuzziness and Knowledge-Based Systems 13(4), 453–464 (2005)
28. Dilworth, R.P., Ward, M.: Residuated lattices. Transactions of the Americal Mathematical Society 45, 335–354 (1939)
29. Dubois, D.: On ignorance and contradiction considered as truth-values. Logic Journal of the IGPL 16(2), 195–216 (2008)
30. Dubois, D., Prade, H.: Can we enforce full compositionality in uncertainty calculi? In: Proceedings of the 11th National Conference on Artificial Intelligence (AAAI 1994), Seattle, Washington, pp. 149–154 (1994)
31. Esteva, F., Garcia-Calvés, P., Godo, L.: Enriched interval bilattices and partial many-valued logics: an approach to deal with graded truth and imprecision. International Journal of Uncertainty, Fuzziness and Knowledge-Based Systems 2(1), 37–54 (1994)
32. Esteva, F., Godo, L.: Monoidal t-norm based logic: towards a logic for left-continuous t-norms. Fuzzy Sets and Systems 124, 271–288 (2001)
33. Gehrke, M., Walker, C.L., Walker, E.A.: Some comments on interval valued fuzzy sets. International Journal of Intelligent Systems 11(10), 751–759 (1996)
34. Gorzałczany, M.B.: A method of inference in approximate reasoning based on interval-valued fuzzy sets. Fuzzy Sets and Systems 21(1), 1–17 (1987)
35. Höhle, U.: Commutative, residuated l-monoids. In: Höhle, U., K.E.P. (eds.) Non-Classical Logics and Their Applications to Fuzzy Subsets: a Handbook of the Mathematical Foundations of Fuzzy Set Theory, pp. 53–106. Kluwer Academic Publishers (1995)
36. Jayaram, B., Rao, C.J.M.: On the distributivity of fuzzy implications over t- and s-norms. IEEE Transactions on Fuzzy Systems 12, 194–198 (2004)
37. Liu, H.W., Wang, G.J.: A note on implicators based on binary aggregation operators in interval-valued fuzzy set theory. Fuzzy Sets and Systems 157, 3231–3236 (2006)
38. Mitchell, H.B.: An intuitionistic OWA operator. International Journal of Uncertainty, Fuzziness and Knowledge-Based Systems 12(6), 843–860 (2004)
39. Moore, R.E.: Methods and applications of interval analysis. Society for Industrial and Applied Mathematics, Philadelphia (1979)
40. Pankowska, A., Wygralak, M.: General IF-sets with triangular norms and their applications to group decision making. Information Sciences 176(18), 2713–2754 (2006)
41. Ruiz-Aguilera, D., Torrens, J.: Distributivity of strong implications over conjunctive and disjunctive uninorms. Kybernetica 42, 319–336 (2005)

42. Ruiz-Aguilera, D., Torrens, J.: Distributivity of residual implications over conjunctive and disjunctive uninorms. Fuzzy Sets and Systems 158, 23–27 (2007)
43. Sambuc, R.: Fonctions Φ-floues. Application à l'aide au diagnostic en pathologie thyroidienne. Ph.D. thesis, Université de Marseille, France (1975)
44. Santiago, R., Bedregal, B., Acióly, B.: Formal aspects of correctness and optimality of interval computations. Formal Aspects Comput. 18, 231–243 (2006)
45. Szmidt, E., Kacprzyk, J.: Distances between intuitionistic fuzzy sets. Fuzzy Sets and Systems 114(3), 505–518 (2000)
46. Trillas, E., Alsina, C.: On the law $[p \wedge q \to r] = [(p \to r) \vee (q \to r)]$ in fuzzy logic. IEEE Transactions on Fuzzy Systems 10, 84–88 (2002)
47. Van Gasse, B.: Interval-valued algebras and fuzzy logics. Ph. D. Thesis, Ghent University, Belgium (2010)
48. Van Gasse, B., Cornelis, C., Deschrijver, G.: Interval-Valued Algebras and Fuzzy Logics. In: Cornelis, C., Deschrijver, G., Nachtegael, M., Schockaert, S., Shi, Y. (eds.) 35 Years of Fuzzy Set Theory. STUDFUZZ, vol. 261, pp. 57–82. Springer, Heidelberg (2010)
49. Van Gasse, B., Cornelis, C., Deschrijver, G., Kerre, E.E.: A characterization of interval-valued residuated lattices. International Journal of Approximate Reasoning 49(2), 478–487 (2008)
50. Van Gasse, B., Cornelis, C., Deschrijver, G., Kerre, E.E.: Triangle algebras: a formal logic approach to interval-valued residuated lattices. Fuzzy Sets and Systems 159(9), 1042–1060 (2008)
51. Van Gasse, B., Cornelis, C., Deschrijver, G., Kerre, E.E.: The pseudo-linear semantics of interval-valued fuzzy logics. Information Sciences 179(6), 717–728 (2009)
52. Van Gasse, B., Deschrijver, G., Cornelis, C., Kerre, E.E.: The standard completeness of interval-valued monoidal t-norm based logic. Information Sciences 189, 63–76 (2012)
53. Vlachos, I.K., Sergiadis, G.D.: Subsethood, entropy, and cardinality for interval-valued fuzzy sets–an algebraic derivation. Fuzzy Sets and Systems 158(12), 1384–1396 (2007)
54. Wei, G.: Some induced geometric aggregation operators with intuitionistic fuzzy information and their application to group decision making. Applied Soft Computing 10, 423–431 (2010)
55. Xu, Z.: Intuitionistic fuzzy aggregation operators. IEEE Transactions on Fuzzy Systems 15, 1179–1187 (2007)
56. Yager, R.R., Rybalov, A.: Uninorm aggregation operators. Fuzzy Sets and Systems 80(1), 111–120 (1996)
57. Zadeh, L.A.: Fuzzy sets. Information and Control 8(3), 338–353 (1965)
58. Zadeh, L.A.: The concept of a linguistic variable and its application to approximate reasoning–I. Information Sciences 8(3), 199–249 (1975)
59. Zalta, E.N.: Basic concepts in modal logic (1995), http://mally.stanford.edu/notes.pdf
60. Zeng, W., Li, H.: Relationship between similarity measure and entropy of interval-valued fuzzy sets. Fuzzy Sets and Systems 157(11), 1477–1484 (2006)

(S,N)-Implications on Bounded Lattices

Benjamín Bedregal, Gleb Beliakov, Humberto Bustince, Javier Fernández,
Ana Pradera, and Renata Reiser

Abstract. Since the birth of the fuzzy sets theory several extensions have been proposed. For these extensions, different sets of membership functions were considered. Since fuzzy connectives, such as conjunctions, negations and implications, play an important role in the theory and applications of fuzzy logics, these connectives have also been extended. An extension of fuzzy logic, which generalizes the ones considered up to the present, was proposed by Joseph Goguen in 1967. In this extension, the membership values are drawn from arbitrary bounded lattices. The simplest and best studied class of fuzzy implications is the class of (S,N)-implications, and in this chapter we provide an extension of (S,N)-implications in the context of bounded lattice valued fuzzy logic, and we show that several properties of this class are preserved in this more general framework.

Benjamín Bedregal
Universidade Federal do Rio Grande do Norte, Departamento de Informática e Matemática
Aplicada, Campus Universitário s/n, Lagoa Nova, 59072-970 Natal-RN, Brazil
e-mail: bedregal@dimap.ufrn.br

Gleb Beliakov
Deakin University, School of Information Technology, Burwood, Australia
e-mail: gleb.beliakov@deakin.edu.au

Humberto Bustince · Javier Fernández
Universidad Pública de Navarra, Departamento de Automática y Computación, Campus
Arrosadia Campus s/n, 31006 Pamplona, Spain
e-mail: bustince@unavarra.es

Ana Pradera
Universidad Rey Juan Carlos, Departamento de Ciencias de la Computación, 28933
Móstoles, Madrid, Spain
e-mail: ana.pradera@urjc.es

Renata Reiser
Universidade Federal de Pelotas, Centro de Desenvolvimento Tecnológico, CP 354, Campus
Universitário, 96010-900 Pelotas-RS, Brazil
e-mail: reiser@inf.ufpel.edu.br

M. Baczyński et al. (Eds.): *Adv. in Fuzzy Implication Functions*, STUDFUZZ 300, pp. 101–124.
DOI: 10.1007/978-3-642-35677-3_5   © Springer-Verlag Berlin Heidelberg 2013

1 Introduction

The necessity of considering truth values other than the classical "true" and "false" was manifested a very long time ago. For example, Plato claimed that between true and false there is a third alternative [35], and Eubulides of Miletus alerts us with the sorites paradox, about the difficulty of determining thresholds for certain properties (or sets), i.e., when an object does possess or not some property. In the modern times, Polish and American logicians Jan Łukasiewicz and Emil Post, respectively, in [45, 54], introduced the idea of 3-valued logics. Later, Kurt Gödel in [30] extended it by considering n possible truth values, and Lotfi Zadeh in [71] introduced the theory of fuzzy sets, where the membership degrees can take values in $[0, 1]$, and therefore, in their logical counterpart, the propositions truth values are real numbers from $[0, 1]$. Since then, several extensions of fuzzy sets have been proposed: for example, interval-valued fuzzy sets [37], Atanassov's intuitionistic fuzzy sets [5], interval-valued intuitionistic fuzzy sets [6], n-dimensional fuzzy sets [58], and some others. These extensions can be considered as special cases of lattice-valued fuzzy sets, introduced by Joseph Goguen in [31], for which the membership degree of an element is an element of a lattice.

An important question in fuzzy logic is that of extending the classical propositional connectives to the fuzzy framework. Since the beginning of fuzzy logics, several particular functions (or families) were proposed to model conjunction, disjunction, negation and implication. The majority of these proposed functions have some common properties, which motivated Claudi Alsina, Enric Trillas and Llorenç Valverde in [4] to use the notion of triangular norm (t-norm in short)[1] and their dual notion (t-conorm) to model conjunction and disjunction in fuzzy logics. In the case of negation, it was Enric Trillas in [64] who proposed the axiomatic accepted nowadays, and in the case of implication, several non-equivalent definitions were proposed (see for example [7, 18, 28, 39, 56, 48, 66, 69]). However, the notion of implication proposed by János Fodor and Marc Roubens in [28], which is equivalent to the notion of [39], has been adopted in several important works on fuzzy implications, such as [9, 18, 48, 60].

In the case of fuzzy implications, several classes of implications obtained from other connectives or functions have been proposed (see for example [9, 14, 26, 49, 59, 70]) and some other works, which extend the usual class of fuzzy implication by substituting the usual connectives, such as t-norms, by a more general operator, for example a uninorm as in [47]. The most natural and best studied class of fuzzy implications is the class of (S, N)-implications, which are based in the well known classical logical equivalence $p \rightarrow q \equiv \neg p \vee q$, where p and q are propositional symbols, $\neg$ denotes negation, $\vee$ denotes disjunction and $\rightarrow$ is the material implication. Enric Trillas and Llorenç Valverde in [65] introduced this class of fuzzy implications, but they only considered continuous t-conorms and strong negations to model fuzzy disjunction and negation, respectively. Later, Alsina and Trillas in [3] introduced

---

[1] T-norms were introduced by Karl Menger in [50] to model distances in probabilistic metric spaces. But it was Berthold Schweizer and Abe Sklar in [57] who gave them the axiomatic form we know today.

the (S,N)-implications as are known nowadays by considering arbitrary t-conorms and negations on the unit interval. When N is strong, the (S,N)-implication is called S-implication.

Since fuzzy connectives (t-norms, t-conorms, negations and implications) play an important role in fuzzy logics, both in theory and applications, several works introduce connectives in each one of the extensions of fuzzy logic (see for example [11, 12, 17, 21, 24, 43]), and in particular for the (S,N) or S-implications (see for example [2, 13, 42]). In the case of lattice-valued fuzzy set theory, there are basically two directions to consider lattice-valued logical connectives:

1. Adding connectives to the lattice structure, in addition to the infimum and supremum operators; i.e. considering enriched lattices with some extra operators, and in general considering properties which relate different connectives. For example, MV-algebras [19] which provide the algebraic setting for Łukasiewicz's infinite-valued propositional logic. MV-algebra does not have an "implication" operator; however there are several algebras in which one of their operators is an "implication" such as the Wajsberg algebra [29], BL-algebra [34], implication algebras [1], MTL-algebras [27], among others (see [68]). For example, BL-algebras are algebras $\langle L, \wedge, \vee, *, \Rightarrow, 0, 1 \rangle$ where $\langle L, \wedge, \vee, 0, 1 \rangle$ is a bounded lattice, $\langle L, *, 1 \rangle$ is a commutative semi-group with 1 as neutral element and $\langle *, \Rightarrow \rangle$ is an adjoint pair, and they satisfy some properties, such as $x \wedge y = x * (x \Rightarrow y)$.

2. Generalizing the notion of fuzzy connective for lattice-valued fuzzy logics by considering the same (or analogous) axioms (conditions) imposed on such connectives. This has been mainly done for aggregation functions (t-norms and t-conorms) and negations (see for example [15, 10, 20, 38, 51, 53, 62]), and for implications on some classes of lattices as, for example, in [44, 38, 46, 55, 63, 67].

In this chapter, we follow the second direction, and we study the natural generalization of (S,N)-implications for (S,N)-implications that take values on arbitrary bounded lattices.

This chapter is organized as follows. In section 2 we provide background on the notions and their main properties, which are necessary to understand the remainder of the text. In particular, we review the notion of bounded lattice, as well as those of negation and t-conorm on bounded lattices. In section 3 we generalize the notion of implication on the unit interval and its main properties, in the context of implications on bounded lattices. In section 4 we define the notion of (S,N)-implication on a bounded lattice, provide a characterization of the subclass of S-implications and present some relations of (S,N)-implication with the properties of implications on the unit interval. In section 5 some final remarks are provided.

2 Preliminaries

In this section we discuss some preliminary notions and properties of bounded lattices, as well as the concepts of negation and t-conorm on bounded lattices. All these concepts are necessary for the remainder of the chapter.

2.1 Bounded Lattices

In this subsection we define some useful concepts on bounded lattices which are based on the papers [15, 53]. If the reader needs a deeper text on lattice theory, we suggest the books [16, 22, 32].

Definition 1. Let $\leq_L$ be a partial order on a set L. The partially ordered set $\langle L, \leq_L \rangle$ is a *lattice* if for all $a, b \in L$ the set $\{a, b\}$ has a supremum and an infimum (in L). If there are two elements, 1 and 0, in L such that $0 \leq_L x \leq_L 1$ for each $x \in L$, then $\langle L, \leq_L, 0, 1 \rangle$ is called a *bounded lattice*.

Definition 2. Let $\wedge$ and $\vee$ be two binary operations on a nonempty set L. Then the algebraic structure $\langle L, \vee, \wedge \rangle$ is a *lattice* if, for each $x, y, z \in L$, the following properties hold:

1. $x \wedge y = y \wedge x$ and $x \vee y = y \vee x$.
2. $(x \wedge y) \wedge z = x \wedge (y \wedge z)$ and $(x \vee y) \vee z = x \vee (y \wedge z)$.
3. $x \wedge (x \vee y) = x$ and $x \vee (x \wedge y) = x$.

If there are elements 1 and 0 in L such that, for all $x \in L$, $x \wedge 1 = x$ and $x \vee 0 = x$, then $\langle L, \vee, \wedge, 0, 1 \rangle$ is a bounded lattice.

Remark 1. It is well known that definitions 1 and 2 are equivalent. Therefore, according to our needs, we shall use one or another. Indeed, if we consider a bounded lattice $\langle L, \leq_L, 0, 1 \rangle$ as a partially ordered set, then the following binary operations: $\forall\, x, y \in L$, $x \wedge y = \inf\{x, y\}$ and $x \vee y = \sup\{x, y\}$ are such that $\langle L, \vee, \wedge, 0, 1 \rangle$ is a bounded lattice in the algebraic sense. Conversely, from a bounded lattice $\langle L, \vee, \wedge, 0, 1 \rangle$ in the algebraic sense, the partial order $\leq_L$ on L defined by $x \leq_L y$ iff $x \vee y = y$ or, equivalently, $x \leq_L y$ iff $x \wedge y = x$, is such that $\langle L, \leq_L, 0, 1 \rangle$ is a bounded lattice.

Let $f, g : L^n \to L$ be n-ary functions. When $f(x_1, \ldots, x_n) \leq_L g(x_1, \ldots, x_n)$ for each $x_1, \ldots, x_n \in L$ then we will denote it by $f \leq_L g$.

Definition 3. Let $\langle L, \leq_L \rangle$ be a lattice. L is said to be *complete* if any $X \subseteq L$ has an infimum and a supremum (in L).

Each complete lattice L is a bounded lattice, in fact $0_L = \inf L$ (or $\sup \emptyset$) and $1_L = \sup L$ (or $\inf \emptyset$).

Definition 4. Let $\langle L, \leq_L, 0_L, 1_L \rangle$ and $\langle M, \leq_M, 0_M, 1_M \rangle$ be bounded lattices. A mapping $f : L \to M$ is said to be a *lattice ord-homomorphism* if, for all $x, y \in L$, it follows that

1. If $x \leq_L y$ then $f(x) \leq_M f(y)$.
2. $f(0_L) = 0_M$ and $f(1_L) = 1_M$.

Remark 2. From now on, we assume that L and M are bounded lattices with the structure $\langle L, \leq_L, 0_L, 1_L \rangle$ and $\langle M, \leq_M, 0_M, 1_M \rangle$, respectively.

Definition 5. A mapping $f : L \longrightarrow M$ is a *lattice alg-homomorphism* if, for all $x, y \in L$, we have

1. $f(x \wedge_L y) = f(x) \wedge_M f(y)$.
2. $f(x \vee_L y) = f(x) \vee_M f(y)$.
3. $f(0_L) = 0_M$ and $f(1_L) = 1_M$.

Proposition 1. *Every alg-homomorphism is an ord-homomorphism, but not every ord-homomorphism is an alg-homomorphism.*

Proof. See page 30 in [33].   □

Definition 6. An ord-homomorphism (alg-homomorphism) $f : L \to M$ is an *ord-isomorphism* (*alg-isomorphism*) if there exists an ord-homomorphism (alg-homomorphism) $f^{-1} : M \to L$ such that $f \circ f^{-1} = Id_M$ and $f^{-1} \circ f = Id_L$, where Id_M (Id_L) is the identity function on M (L). f^{-1} is said to be the *inverse* of f.

Contrary to the case of ord-homomorphism and alg-homomorphism, both notions of isomorphism agree, in the sense that f is an ord-isomorphism if and only if (iff) f is an alg-isomorphism. Therefore, we will simply call isomorphisms to ord-isomorphisms and alg-isomorphisms.

Proposition 2. *A function $f : L \to M$ is an isomorphism iff f is bijective and for each $x, y \in L$, we have that*

$$x \leq_L y \text{ iff } f(x) \leq_M f(y). \tag{1}$$

Proof. ($\Rightarrow$) By definition, f is trivially bijective. Since it is an ord-homomorphism, it remains to prove that $f(x) \leq_M f(y)$ implies that $x \leq_L y$. Since, f^{-1} is an isomorphism then $f(x) \leq_M f(y)$ implies that $f^{-1}(f(x)) \leq_L f^{-1}(f(y))$ and so $x \leq_L y$.

($\Leftarrow$) Since f is bijective, then it has an inverse f^{-1} and so $f \circ f^{-1} = Id_M$ and $f^{-1} \circ f = Id_L$. On the other hand, by equation (1), f is and ord-homomorphism. Let $x, y \in M$ be such that $x \leq_M y$. Then, as f is bijective, there exist $a, b \in L$ such that (*) $f(a) = x$ and $f(b) = y$ and so $f(a) \leq_M f(b)$. Hence, by equation (1), $a \leq_L b$. But, by (*), $a = f^{-1}(x)$ and $b = f^{-1}(y)$ and therefore, $f^{-1}(x) \leq_L f^{-1}(y)$.   □

When L and M are the same lattice, we say that an isomorphism is an *automorphism*, or *L-automorphism* when it is important to refer to a specific lattice. We will denote the set of all automorphisms on a bounded lattice L by $Aut(L)$. Since Id_L is clearly an automorphism and automorphisms are closed under composition and inversion, then the algebra $\langle Aut(L), \circ \rangle$ is a group. In algebra, an important tool is the action of groups on sets [36]. In our case, the action of the automorphism group transforms lattice functions on other lattice functions.

Definition 7. Given a function $f : L^n \to L$, the *action of an L-autormorphism ρ over f* is the function $f^\rho : L^n \to L$ defined as in equation (2).

$$f^\rho(x_1, \ldots, x_n) = \rho^{-1}(f(\rho(x_1), \ldots, \rho(x_n))). \tag{2}$$

f^ρ is said to be a *conjugate* of f.

Notice that if $f : L^n \to L$ is a conjugate of $g : L^n \to L$ and g is a conjugate of $h : L^n \to L$ then, because L-automorphisms are closed under composition, f is a conjugate of h; and if f is a conjugate of g then, because the inverse of an L-automorphism is also an L-automorphism, g is also a conjugate of f. Thus, the automorphism action on the set of n-ary functions on L (L^{L^n}) determines an equivalence relation on L^{L^n}.

2.2 Negations on L

Fuzzy negations are generalizations of the classical negation $\neg$ and, as in classical logics, they have been used to define other connectives from binary connectives. In this subsection we present a natural extension of negations on the unit interval by considering arbitrary bounded lattices as possible sets of truth values.

Definition 8. A mapping $N : L \to L$ is a *negation on L*, if the following properties are satisfied for each $x, y \in L$:

(N1) $N(0_L) = 1_L$ and $N(1_L) = 0_L$.
(N2)   If $x \leq_L y$ then $N(y) \leq_L N(x)$.

Moreover, a negation N on L is *strong* if it also satisfies the involution property, i.e.

(N3) $N(N(x)) = x$ for each $x \in L$.

A negation N on L is called *frontier* if it satisfies the property:

(N4) $N(x) \in \{0_L, 1_L\}$ iff $x = 0_L$ or $x = 1_L$.

Observe that each strong negation on L is a frontier negation on L and that a negation on L is frontier iff it is both non-filling ($N(x) = 1_L$ iff $x = 0_L$) and non-vanishing ($N(x) = 0_L$ iff $x = 1_L$) (see [9], pg. 14).

Proposition 3. *The functions $N_\perp, N_\top : L \to L$ defined by*

$$N_\perp(x) = \begin{cases} 1_L & \text{if } x = 0_L, \\ 0_L & \text{otherwise,} \end{cases}$$

and

$$N_\top(x) = \begin{cases} 0_L & \text{if } x = 1_L, \\ 1_L & \text{otherwise,} \end{cases}$$

are negations on L, such that for any negation N on L, we have that $N_\perp \leq_L N \leq_L N_\top$.

Proof. Straightforward.   □

An element $e \in L$ is an equilibrium point of a negation N on L if $N(e) = e$.

Proposition 4. *Let $N : L \to L$, ρ be an L-automorphism and $i \in \{1, \ldots, 4\}$. N satisfies (Ni) iff N^ρ satisfies (Ni). Moreover, e is an equilibrium point of N iff $\rho^{-1}(e)$ is an equilibrium point of N^ρ.*

Proof. ($\Rightarrow$)

(N1) $N^{\rho}(0_L) = \rho^{-1}(N(\rho(0_L))) = \rho^{-1}(N(0_L)) = \rho^{-1}(1_L) = 1_L$. Analogously, $N^{\rho}(1_L) = 0_L$.

(N2) If $x \leq_L y$ then $\rho(x) \leq_L \rho(y)$ and so $N(\rho(y)) \leq_L N(\rho(x))$. Therefore, by isotonicity of ρ^{-1}, $\rho^{-1}(N(\rho(y))) \leq_L \rho^{-1}(N(\rho(x)))$.

(N3) $N^{\rho}(N^{\rho}(x)) = \rho^{-1}(N(\rho(\rho^{-1}(N(\rho(x)))))) = \rho^{-1}(N(N(\rho(x)))) = \rho^{-1}(\rho(x)) = x$.

(N4) If $N^{\rho}(x) = 0_L$ then, by eq. (2) and because $\rho(0_L) = 0_L$, $N(\rho(x)) = 0_L$. So, because N satisfies (N4), $\rho(x) \in \{0_L, 1_L\}$ and therefore $x = 0_L$ or $x = 1_L$. Analogously for $N^{\rho}(x) = 1_L$.

Moreover, $N(e) = e$ iff $N(\rho(\rho^{-1}(e))) = e$. So, since ρ^{-1} is bijective, $N(e) = e$ iff $N^{\rho}(\rho^{-1}(e)) = \rho^{-1}(e)$.

($\Leftarrow$) Straightforward from the previous item and the fact that for any function $f : L \to L$, $(f^{\rho})^{\rho^{-1}} = f$. □

Corollary 1. *Let $N : L \to L$ and ρ be an L-automorphism. N is a (strong, frontier) negation on L iff N^{ρ} is a (strong, frontier) negation on L.*

Proof. Straightforward from Proposition 4. □

2.3 T-Conorms on L

In fuzzy logics, classical disjunctions have been modeled via functions called t-conorms. In this subsection we present the natural generalization of this concept for arbitrary bounded lattices as possible sets of truth values.

Definition 9. A mapping $S : L \times L \to L$ is a *t-conorm* on L if the following properties are satisfied for each $x, y, z \in L$:

(S1) $S(x,y) = S(y,x)$.
(S2) $S(x, S(y,z)) = S(S(x,y), z)$.
(S3) if $y \leq_L z$ then $S(x,y) \leq_L S(x,z)$.
(S4) $S(x, 0_L) = x$.

S is *positive* if for each $x, y \in L$ it satisfies the property

(S5) $S(x,y) = 1_L$ iff $x = 1_L$ or $y = 1_L$.

$x \in L$ is an *idempotent element* of S if $S(x,x) = x$.

Lemma 1. *Let S be a t-conorm on L. Then for each $x, y \in L$, $x \vee_L y \leq_L S(x,y)$.*

Proof. By (S4) and (S3), $x = S(x, 0_L) \leq_L S(x,y)$. Analogously, $y \leq_L S(x,y)$ and therefore, $x \vee_L y \leq_L S(x,y)$. □

Proposition 5. *Let S a t-conorm on L. Then $S_{\perp} \leq_L S \leq_L S_{\top}$, where*

$$S_{\perp}(x,y) = x \vee_L y \text{ and } S_{\top}(x,y) = \begin{cases} x \vee_L y & \text{if } x = 0_L \text{ or } y = 0_L, \\ 1_L & \text{otherwise.} \end{cases} \quad (3)$$

Proof. Clearly, $S_\perp$ and $S_\top$ are t-conorms on L. By Lemma 1, $S_\perp \leq S$ and trivially, $S \leq_L S_\top$. □

Proposition 6. *Let S be a t-conorm on L. Each $x \in L$ is an idempotent element of S iff $S = S_\perp$.*

Proof. ($\Rightarrow$) Let $x, y \in L$. Then by Lemma 1, $x \vee_L y \leq_L S(x, y)$ and so because each element of L is idempotent of S, then (*) $S(x \vee_L y, x \vee_L y) \leq_L S(x, y)$. On the other hand, since $x \leq_L x \vee_L y$ and $y \leq_L x \vee_L y$, then by (S3), $S(x, y) \leq_L S(x \vee_L y, x \vee_L y)$. So, by (*), $S(x, y) = S(x \vee_L y, x \vee_L y)$ and therefore, $S(x, y) = x \vee_L y$, i.e. $S = S_\perp$.
($\Leftarrow$) Straightforward from equation (3). □

Proposition 7. *Let $S : L \times L \to L$, ρ be an L-automorphism and $i \in \{1, \ldots, 5\}$. S satisfies (Si) iff S^ρ satisfies (Si).*

Proof. ($\Rightarrow$) Let $x, y, z \in L$. Then

(S1)   Straightforward.
(S2)
$$\begin{aligned}
S^\rho(x, S^\rho(y, z)) &= \rho^{-1}(S(\rho(x), \rho(\rho^{-1}(S(\rho(y), \rho(z)))))) \quad \text{by eq. (2)} \\
&= \rho^{-1}(S(\rho(x), S(\rho(y), \rho(z)))) \\
&= \rho^{-1}(S(S(\rho(x), \rho(y)), \rho(z))) \quad\quad \text{by (S2)} \\
&= \rho^{-1}(S(\rho(\rho^{-1}(S(\rho(x), \rho(y)))), \rho(z))) \\
&= S^\rho(S^\rho(x, y), z) \quad\quad\quad\quad \text{by eq. (2).}
\end{aligned}$$
(S3)   If $y \leq_L z$ then $\rho(y) \leq_L \rho(z)$ and so, by (S3), $S(\rho(x), \rho(y)) \leq_L S(\rho(x), \rho(z))$. Therefore, $\rho^{-1}(S(\rho(x), \rho(y))) \leq_L \rho^{-1}(S(\rho(x), \rho(z)))$, i.e. $S^\rho(x, y) \leq_L S^\rho(x, z)$.
(S4) $S^\rho(x, 0_L) = \rho^{-1}(S(\rho(x), \rho(0_L))) = \rho^{-1}(S(\rho(x), 0_L)) = \rho^{-1}(\rho(x)) = x$.
(S5) $S^\rho(x, y) = 1_L$ iff $\rho^{-1}(S(\rho(x), \rho(y))) = 1_L$ iff $S(\rho(x), \rho(y)) = 1_L$ iff, by (S5), $\rho(x) = 1_L$ or $\rho(y) = 1_L$ iff $x = 1_L$ or $y = 1_L$.

($\Leftarrow$) Straightforward from the previous item and the fact that for any function $f : L^n \to L$, $(f^\rho)^{\rho^{-1}} = f$. □

Corollary 2. *Let $S : L \times L \to L$ and ρ be an L-automorphism. S is a t-conorm on L iff S^ρ is a t-conorm on L. Moreover, S is positive iff S^ρ also is positive.*

Proof. Straightforward from Proposition 7. □

In [41, 52] it was observed that it is possible to obtain, in a canonical way, a negation N_S on the unit interval from a t-conorm S called the natural negation of S or the negation induced by S. In the general case, where we have a t-conorm on a bounded lattice L, it is not always possible, because the construction of N_S is based on the infimum, of possibly, an infinite number of elements.

Proposition 8. *Let L be a complete lattice and S be a t-conorm on L. Then the function $N_S : L \to L$ defined by*

$$N_S(x) = \inf\{z \in L : S(x, z) = 1_L\} \tag{4}$$

is a negation on L.

Proof. (N1) $N_S(1_L) = \inf\{z \in L : S(1_L, z) = 1_L\} = \inf L = 0_L$ and $N_S(0_L) = \inf\{z \in L : S(0_L, z) = 1_L\} = \inf\{1_L\} = 1_L$.

(N2)   If $x \leq_L y$ then for any $z \in L$, $S(x,z) \leq_L S(y,z)$ and therefore, if $S(x,z) = 1_L$ then $S(y,z) = 1_L$. So, $\{z \in L : S(x,z) = 1_L\} \subseteq \{z \in L : S(y,z) = 1_L\}$. Hence, $N_S(y) = \inf\{z \in L : S(y,z) = 1_L\} \leq_L \inf\{z \in L : S(x,z) = 1_L\} = N_S(x)$.   □

Lemma 2. *Let L be a complete lattice and S be a t-conorm on L. If S is positive then $N_S = N_\top$.*

Proof. If $x \neq 1_L$ and $z \in L$ then, by (S5), $S(x,z) = 1_L$ iff $z = 1_L$. So, by eq. (4), $N_S(x) = \inf\{1_L\} = 1_L$. Therefore, $N_S = N_\top$.   □

Proposition 9. *Let L be a complete lattice, S be a t-conorm on L and ρ be an L-automorphism. Then $N_S^\rho = N_{S^\rho}$.*

Proof. Let $x \in L$, then $N_S^\rho(x) = \rho^{-1}(N_S(\rho(x))) = \rho^{-1}(\inf\{z \in L : S(\rho(x), z) = 1_L\}) = \rho^{-1}(\inf\{z \in L : S^\rho(x, \rho^{-1}(z)) = 1_L\}) = \inf\{\rho^{-1}(z) \in L : S^\rho(x, \rho^{-1}(z)) = 1_L\} = \inf\{z \in L : S^\rho(x, z) = 1_L\} = N_{S^\rho}(x)$.   □

3   Implications on L

3.1   Basic Definitions and Notations

In the literature several definitions of fuzzy implications are considered, see for example [7, 18, 28, 39, 56, 48, 66, 69]. Here we will consider the definition of [28] which is equivalent to the definition of [39], and that was adopted in several important works, such as [9, 18, 48, 60]. Here we adapt this definition for the context of implications on bounded lattices.

Definition 10. A function $I : L \times L \to L$ is an implication on L if for each $x, y, z \in L$ we have that

1. First place antitonicity (FPA): if $x \leq_L y$ then $I(y,z) \leq_L I(x,z)$.
2. Second place isotonicity (SPI): if $y \leq_L z$ then $I(x,y) \leq_L I(x,z)$.
3. Corner condition 1 (CC1): $I(0_L, 0_L) = 1_L$.
4. Corner condition 2 (CC2): $I(1_L, 1_L) = 1_L$.
5. Corner condition 3 (CC3): $I(1_L, 0_L) = 0_L$.

The set of implications on L will be denoted by $\mathscr{I}_L$.

Notice that from (SPI) and (CC1) it is possible to deduce the left boundary condition (LB) $I(0_L, y) = 1_L$ and from (FPA) and (CC2) the right boundary condition (RB) $I(x, 1_L) = 1_L$ and so the last corner condition (CC4) $I(0_L, 1_L) = 1_L$.

Given two implications on L, I_1 and I_2, we say that I_1 is smaller than I_2, denoted by $I_1 \leq_L I_2$ if for each $x, y \in L$, $I_1(x,y) \leq_L I_2(x,y)$. It is clear that $\leq_L$ is a partial order on $\mathscr{I}_L$.

Theorem 1. *For each bounded lattice* L, $\langle \mathscr{I}_L, \leq_L, I_\perp, I_\top \rangle$, *where* $I_\perp, I_\top : L \times L \to L$ *defined by*

$$I_\perp(x,y) = \begin{cases} 1_L & \text{if } x = 0_L \text{ or } y = 1_L, \\ 0_L & \text{otherwise,} \end{cases}$$

$$I_\top(x,y) = \begin{cases} 0_L & \text{if } x = 1_L \text{ and } y = 0_L, \\ 1_L & \text{otherwise,} \end{cases}$$

for any $x, y \in L$, *is a bounded lattice.*

Proof. It is not hard to prove that given two implications on L, I_1 and I_2, the functions $I_\vee, I_\wedge : L \times L \to L$ defined by

$$I_\vee(x,y) = I_1(x,y) \vee_L I_2(x,y) \text{ and } I_\wedge(x,y) = I_1(x,y) \wedge_L I_2(x,y) \tag{5}$$

are implications on L. For example, for the case of (FPA), let $x, y, z \in L$ be such that $x \leq_L y$. Then by (FPA) $I_1(y,z) \leq_L I_1(x,z)$ and $I_2(y,z) \leq_L I_2(x,z)$. So, $I_1(y,z) \vee_L I_2(y,z) \leq_L I_1(x,z) \vee_L I_2(x,z)$. Therefore, by equation (5), $I_\vee(y,z) \leq_L I_\vee(x,z)$ which means that $I_\vee$ satisfies (FPA).

However we also need to prove that $I_\vee$ and $I_\wedge$ are the supremum and infimum, respectively, of I_1 and I_2 with respect to $\leq_L$. Suppose that I is an implication on L which is an upper bound of I_1 and I_2. Then, for any $x, y \in L$, $I_1(x,y) \leq_L I(x,y)$ and $I_2(x,y) \leq_L I(x,y)$. So, $I_1(x,y) \vee_L I_2(x,y) \leq_L I(x,y)$. Therefore, by equation (5), $I_\vee(x,y) \leq_L I(x,y)$ and hence $I_\vee \leq_L I$. Thus, since trivially, $I_1 \leq_L I_\vee$ and $I_2 \leq_L I_\vee$, then $I_\vee$ is the minimal upper bound of I_1 and I_2 with respect to $\leq$, i.e. $I_\vee = I_1 \vee_{\mathscr{I}_L} I_2$.

Therefore, since $I_\perp$ and $I_\top$ are clearly implications on L and for any other implication I on L we have that $I_\perp \leq_L I \leq_L I_\top$, then $\langle \mathscr{I}_L, \leq_L, I_\perp, I_\top \rangle$ is a bounded lattice. $\square$

Proposition 10. *Let* ρ *be an L-automorphism,* $I : L \times L \to L$ *be a function and* $P \in \{(FPA), (SPI), (CC1), (CC2), (CC3), (CC4), (LB), (RB)\}$. *$I$ satisfies P iff I^ρ also satisfies P.*

Proof. ($\Rightarrow$)

- (FPA) Let $x, y, z \in L$ such that $x \leq_L y$, then by isotonicity of L-automorphisms, $\rho(x) \leq_L \rho(y)$ and so, by (FPA), $I(\rho(y), \rho(z)) \leq_L I(\rho(x), \rho(z))$. Therefore, since ρ^{-1} is also an L-automorphism and by isotonicity of L-automorphisms, we have that $\rho^{-1}(I(\rho(y), \rho(z))) \leq_L \rho^{-1}(I(\rho(x), \rho(z)))$. Hence, $I^\rho(y,z) \leq_L I^\rho(x,z)$, i.e. I^ρ satisfies (FPA).
- (SPI) Analogous to (FPA).
- (CC1), (CC2), (CC3) and (CC4) Since $\rho(0_L) = 0_L$ and $\rho(1_L) = 1_L$ then it is straightforward that I^ρ satisfies the corner conditions.
- (LB) $I^\rho(0_L, y) = \rho^{-1}(I(\rho(0_L), \rho(y))) = \rho^{-1}(I(0_L, \rho(y))) = \rho^{-1}(\rho(y)) = y$.
- (RB) Analogous to (LB).

($\Leftarrow$) Straightforward taking into account the previous item and the fact that for any function $f : L^n \to L$, $(f^\rho)^{\rho^{-1}} = f$. $\square$

Corollary 3. *Let ρ be an L-automorphism. Then, for any implication I on L, I^ρ is also an implication on L. In particular, $I_\perp^\rho = I_\perp$ and $I_\top^\rho = I_\top$.*

Proof. Straightforward because L-automorphisms are bijections, so $\rho(x) = 0_L$ iff $x = 0_L$, and, $\rho(x) = 1_L$ iff $x = 1_L$. Therefore, $I_\perp^\rho = I_\perp$ and $I_\top^\rho = I_\top$. $\square$

In the following we will provide a relaxed version of Proposition 1 of [18].

Proposition 11. *Consider a function $I : L \times L \to L$. Then*

1. I satisfies (FPA) iff for any $x,y,z \in L$, $I(x \vee_L y, z) \leq_L I(x,z) \wedge_L I(y,z)$.
2. I satisfies (FPA) iff for any $x,y,z \in L$, $I(x,z) \vee_L I(y,z) \leq_l I(x \wedge_L y, z)$.
3. I satisfies (SPI) iff for any $x,y,z \in L$, $I(x, y \wedge_L z) \leq_L I(x,y) \wedge_L I(x,z)$.
4. I satisfies (SPI) iff for any $x,y,z \in L$, $I(x,y) \vee_L I(x,z) \leq_L I(x, y \vee_L z)$.

Proof. 1. ($\Rightarrow$) Since $x \leq_L x \vee_L y$ and $y \leq_L x \vee_L y$ then by (FPA), $I(x \vee_L y, z) \leq_L I(x,z)$ and $I(x \vee_L y, z) \leq_L I(y,z)$. Therefore, $I(x \vee_L y, z) \leq_L I(x,z) \wedge_L I(y,z)$.
($\Leftarrow$) If $x \leq_L y$ then $x \vee_L y = y$ and so $I(x \vee_L y, z) = I(y,z)$. But by hypothesis, $I(x \vee_L y, z) \leq_L I(x,z) \wedge_L I(y,z)$. Therefore, $I(y,z) \leq_L I(x,z) \wedge_L I(y,z)$ which implies that $I(x,z) \wedge_L I(y,z) = I(y,z)$. Hence, $I(y,z) \leq_L I(x,z)$.
The rest of the items are proved in a similar way. $\square$

3.2 Extra Properties

Apart from properties (FPA), (SPI) and the corner conditions, several other properties of implications on the unit interval have been considered. These properties and their interdependencies, as well as some other independent properties can be found in [9, 18, 60, 61]. Here we adapt some of these properties for the context of implications on bounded lattices.

Definition 11. An implication I on L satisfies

1. the left neutrality property (NP) if for each $y \in L$, $I(1_L, y) = y$.
2. the exchange principle (EP) if for each $x,y,z \in L$, $I(x, I(y,z)) = I(y, I(x,z))$.
3. the identity principle (IP) if for each $x \in L$, $I(x,x) = 1_L$.
4. the ordering property (OP) if for each $x,y \in L$, $I(x,y) = 1_L$ iff $x \leq_L y$.
5. the iterative Boolean law (IBL) if for each $x,y \in L$, $I(x, I(x,y)) = I(x,y)$.
6. the consequent boundary (CB) if for each $x,y \in L$, $y \leq_L I(x,y)$.

Lemma 3. *Let $I : L \times L \to L$ be a function.*

1. if I satisfies (LB) then I satisfies (CC1) and (CC4).
2. if I satisfies (RB) then I satisfies (CC2) and (CC4).
3. if I satisfies (NP) then I satisfies (CC2) and (CC3).
4. if I satisfies (IP) then I satisfies (CC1) and (CC2).
5. if I satisfies (OP) then I satisfies (CC1), (CC2), (CC4), (LB), (RB) and (IP).

Proof. Straightforward. $\square$

Lemma 4. *If a function $I : L \times L \to L$ satisfies (OP) and (EP) then I satisfies (FPA), (CC1), (CC2), (CC3), (CC4), (LB), (RB), (NP) and (IP).*

Proof. Analogous to the proof of Lemma 1.3.4 in [9]. $\square$

Proposition 12. *Let ρ be an L-automorphism and $I : L \times L \to L$ be a function. I satisfies a property $P \in \{(NP),(EP),(IP),(OP),(IBL),(CB)\}$ iff I^ρ also satisfies P.*

Proof. $(\Rightarrow)$

- (NP) $I^\rho(1_L, y) = \rho^{-1}(I(\rho(1_L), \rho(y))) = \rho^{-1}(I(1_L, \rho(y))) = \rho^{-1}(\rho(y)) = y.$
- (EP)
  $$
  \begin{aligned}
  I^\rho(x, I^\rho(y,z)) &= \rho^{-1}(I(\rho(x), \rho(\rho^{-1}(I(\rho(y), \rho(z)))))) \quad \text{by eq. (2)} \\
  &= \rho^{-1}(I(\rho(x), I(\rho(y), \rho(z)))) \\
  &= \rho^{-1}(I(\rho(y), I(\rho(x), \rho(z)))) \qquad\qquad \text{by (EP)} \\
  &= \rho^{-1}(I(\rho(y), \rho(\rho^{-1}(I(\rho(x), \rho(z)))))) \\
  &= I^\rho(y, I^\rho(x,z)) \qquad\qquad\qquad\qquad \text{by eq. (2).}
  \end{aligned}
  $$
- (IP) $I^\rho(x,x) = \rho^{-1}(I(\rho(x), \rho(x))) = \rho^{-1}(1_L) = 1_L.$
- (OP) On one hand, if $I^\rho(x,y) = 1_L$ then, by eq. (2), $\rho^{-1}(I(\rho(x), \rho(y))) = 1_L$ and so, $I(\rho(x), \rho(y)) = 1_L$. Thus, as I satisfies (OP), $\rho(x) \leq_L \rho(y)$. Therefore, by equation (1), $x \leq_L y$. On the other hand, if $x \leq_L y$, then, by equation (1), $\rho(x) \leq_L \rho(y)$. Therefore, as I satisfies (OP), $I(\rho(x), \rho(y)) = 1_L$ and so $\rho^{-1}(I(\rho(x), \rho(y))) = \rho^{-1}(1_L) = 1_L$, i.e. $I^\rho(x,y) = 1_L$.
- (IBL)
  $$
  \begin{aligned}
  I^\rho(x, I^\rho(x,y)) &= \rho^{-1}(I(\rho(x), \rho(\rho^{-1}(I(\rho(x), \rho(y)))))) \quad \text{by eq. (2)} \\
  &= \rho^{-1}(I(\rho(x), I(\rho(x), \rho(y)))) \\
  &= \rho^{-1}(I(\rho(x), \rho(y))) \qquad\qquad\qquad \text{by (IBL)} \\
  &= I^\rho(x,y) \qquad\qquad\qquad\qquad\quad \text{by eq. (2).}
  \end{aligned}
  $$
- (CB) Since I satisfies (CB), then for any $x,y \in L$, $\rho(y) \leq_L I(\rho(x), \rho(y))$. So, by equations (2) and (1), $y = \rho^{-1}(\rho(y)) \leq_L \rho^{-1}(I(\rho(x), \rho(y))) = I^\rho(x,y).$

$(\Leftarrow)$ Straightforward from the previous item and the fact that for any function $f : L^n \to L$, $(f^\rho)^{\rho^{-1}} = f$. $\square$

3.3   Relation between Negations and Implications on L

As it is well known in classic logic, the negation of a propositional formula p is logically equivalent to $p \to f$ where f denotes the absurd, a contradiction and so its truth value in any interpretation is always "false". In the following we generalize this dependency of the negation on the implication for negations and implications on L.

Proposition 13. *If a function $I : L \times L \to L$ satisfies (FPA), (CC1) and (CC3) then the function $N_I : L \to L$ defined for each $x \in L$ by*

$$
N_I(x) = I(x, 0_L) \tag{6}
$$

is a negation on L.

Proof. By (CC1), $N_I(0_L) = I(0_L, 0_L) = 1_L$ and by (CC3), $N_I(1_L) = I(1_L, 0_L) = 0_L$ and so N_I satisfies (N1). If $x \leq_L y$ then by (FPA), $I(y, 0) \leq_L I(x, 0)$ and therefore, $N_I(y) \leq_L N_I(x)$, i.e. N_I satisfies (N2). $\qquad\square$

Thus, for each implication I on L, N_I is a negation on L called the negation on L induced by I. In particular when N_I is strong we say that I satisfies the property (SNI), and if N_I is a frontier negation on L, then we say that I satisfies the property (FNI).

Proposition 14. *If a function $I : L \times L \to L$ satisfies (EP) and (OP) then N_I is a negation on L, $x \leq_L N_I(N_I(x))$ for each $x \in L$ and $N_I \circ N_I \circ N_I = N_I$.*

Proof. Analogous to Proposition 1.4.17 in [9]. $\qquad\square$

Proposition 15. *Let I be an implication on L and ρ be an L-automorphism. Then $(N_I)^\rho = N_{I^\rho}$.*

Proof. Analogous to Theorem 1.4.21 of [9]. $\qquad\square$

3.4 Contraposition

Definition 12. Let $I : L \times L \to L$ and N be a negation on L. I satisfies

1. contrapositivity property (CP) for N if for each $x, y \in L$, $I(x, y) = I(N(y), N(x))$.
2. left contrapositivity property (LCP) for N if for each $x, y \in L$, $I(N(x), y) = I(N(y), x)$.
3. right contrapositivity property (RCP) for N if for each $x, y \in L$, $I(x, N(y)) = I(y, N(x))$.

Remark 3. If N is strong, then I satisfies (CP) for N iff I satisfies (LCP) for N iff I satisfies (RCP) for N.

Proposition 16. *Let $I : L \times L \to L$, N be a negation on L and ρ an L-automorphism. I satisfies (CP), (LCP), or (RCP) for N iff I^ρ satisfies (CP), (LCP), or (RCP) for N^ρ.*

Proof. We will prove just the case of (LCP) because the cases of (CP) and (RCP) are analogous.

($\Rightarrow$) For each $x, y \in L$, we have that
$$\begin{aligned} I^\rho(N^\rho(x), y) &= \rho^{-1}(I(\rho(\rho^{-1}(N(\rho(x)))), \rho(y))) && \text{by eq. (2)} \\ &= \rho^{-1}(I(N(\rho(y)), \rho(x))) && \text{by (LCP)} \\ &= \rho^{-1}(I(\rho(\rho^{-1}(N(\rho(y)))), \rho(x))) && \\ &= I^\rho(N^\rho(y), x) && \text{by eq. (2).} \end{aligned}$$
So, I^ρ satisfies (LCP) for N^ρ.

($\Leftarrow$) Straightforward from the previous item since for each $f : L^n \to L$, $(f^\rho)^{\rho^{-1}} = f$. $\qquad\square$

Proposition 17. *Let N be a negation on L and $I : L \times L \to L$. If I satisfies (NP) and (LCP) for N then*

1. *I is an implication on L iff it satisfies (FPA) and (SPI).*
2. $N_I \circ N = Id_L$.

Proof. Analogous to Lemma 1.5.14 in [9]. □

The following result is analogous to Proposition 3.6 in [60], but including the property (IBL) and relaxing the requirement of considering only strong negations.

Proposition 18. *For any bounded lattice L with at least three elements and a frontier negation N on L, the greatest implication on L, i.e. $I_\top$, satisfies (CB), (IP), (EP), (IBL), (CP) with respect to N but not (NP).*

Proof. Let $I_\top$ be the greatest implication on L (see Theorem 1).

Trivially, for any $x \in L$, $I_\top(x,x) = 1_L$.

If $x = 1_L$ and $y = 0_L$, then $y \leq_L I_\top(x,y)$, $I_\top(x,I_\top(y,z)) = I_\top(x,I_\top(0_L,z)) = I_\top(x,1_L) = 1_L = I_\top(0_L,I_\top(x,z)) = I_\top(y,I_\top(x,z))$, $I_\top(x,I_\top(x,y)) = I_\top(x,I_\top(1_L, 0_L)) = I_\top(x,0_L) = I_\top(x,y)$, and $I_\top(x,y) = I_\top(1_L,0_L) = I_\top(N(0_L),N(1_L)) = I_\top(N(y), N(x))$.

If $x = 1_L$ and $y = 1_L$, then $y \leq_L 1_L = I_\top(x,y)$, $I_\top(x,I_\top(y,z)) = I_\top(y,I_\top(x,z))$, $I_\top(x,I_\top(x,y)) = I_\top(x,I_\top(1_L,1_L)) = I_\top(x,1_L) = I_\top(x,y)$, and $I_\top(x,y) = I_\top(1_L,1_L) = 1_L = I_\top(0_L,0_L) = I_\top(N(1_L),N(1_L)) = I_\top(N(y),N(x))$.

If $x = 1_L$ and $y \in L - \{0_L,1_L\}$, then $y \leq_L 1_L = I_\top(x,y)$, $I_\top(x,I_\top(y,z)) = I_\top(x,1_L) = 1_L = I_\top(y,I_\top(x,z))$, $I_\top(x,I_\top(x,y)) = I_\top(x,1_L) = 1_L = I_\top(x,y)$, and since N is frontier $N(y) \neq 1_L$, $I_\top(x,y) = 1_L = I_\top(N(y),N(x))$.

If $x \neq 1_L$ then for each $y \in L$, $y \leq_L 1_L = I_\top(x,y)$, $I_\top(x,I_\top(y,z)) = 1_L = I_\top(y,1_L) = I_\top(y,I_\top(x,z))$, $I_\top(x,I_\top(x,y)) = 1_L = I_\top(x,y)$, and since N is frontier $N(x) \neq 0_L$, $I_\top(x,y) = 1_L = I_\top(N(y),N(x))$.

Therefore, for each $x,y,z \in L$, $y \leq_L I_\top(x,y)$, $I_\top(x,x) = 1_L$, $I_\top(x,I_\top(y,z)) = I_\top(y,I_\top(x,z))$, $I_\top(x,I_\top(x,y)) = I_\top(x,y)$ and for each frontier negation N on L, $I_\top(x,y) = I_\top(N(y),N(x))$. So, $I_\top$ satisfies (CB), (IP), (EP), (IBL), (CP) with respect to N. Nevertheless, for any $x \in L - \{0_L,1_L\}$, $I_\top(1_L,x) = 1_L \neq x$ and therefore it does not satisfy (NP). □

Proposition 19. *Let $I : L \times L \to L$ be a function. Then*

1. *if I satisfies (NP) and (CP) for some negation N on L then N is strong and $N = N_I$.*
2. *if N_I is a strong negation on L and I satisfies (CP) with respect to N_I then I satisfies (NP).*
3. *if I satisfies (SPI) and N_I is a strong negation on L then I satisfies (LB).*
4. *if I satisfies (EP) and N_I is a strong negation on L then I satisfies (CC1), (CC2), (CC3), (NP), and (CP) only with respect to N_I.*
5. *if I satisfies (FPA), (NP) and (CP) for some negation N on L then N is strong, $N_I = N$ and I satisfies (SPI), (LB), (RB), (CC3) and (CB).*

Proof. Item 1 is analogous to Lemma 1.5.4.-(v) in [9]. Item 2 is analogous to Lemma 1.5.6.-(i) in [9]. Item 3 is analogous to Lemma 1 in [18]. Item 4 is analogous to Lemma 1.5.6.-(ii) in [9]. Item 5 is a mix between Lemma 1 in [18] and Lemma 1.5.6.-(i) in [9]. □

The next lemma is the L-valued version to Lemma 2 in [18], but here we only require that N_I is frontier instead of strong, and we abolish the necessity of (NP) in the last item.

Lemma 5. *Let $I : L \times L \to L$. If I satisfies*

1. (SPI), (NP) and (FNI); or
2. (FPA), (NP) and (FNI); or
3. (RB), (EP), (FNI) and (IP); or
4. (EP), (FNI) and (*) $I(x,x) = I(0_L,x)$ for each $x \in L$,

then for each $x,y \in L$

$$I(x,y) = 0_L \text{ iff } x = 1_L \text{ and } y = 0_L$$

Proof. ($\Rightarrow$) Suppose that (**) $I(x,y) = 0_L$

1. From (**), by (SPI), we have that $I(x,0_L) = 0_L$, i.e. $N_I(x) = 0_L$. So, by (FNI), $x = 1_L$. Therefore, by (**), $I(1_L,y) = 0_L$ and so by (NP), $y = 0_L$.
2. From (**), by (FPA), we have that $I(1_L,y) = 0_L$, and so by (NP), $y = 0_L$. Therefore, by (**), $I(x,0_L) = 0_L$, i.e. $N_I(x) = 0_L$. Hence, by (FNI), $x = 1_L$.
3.
$$
\begin{aligned}
N_I(y) &= I(y,0_L) && \text{by eq. (6)} \\
&= I(y,I(x,y)) && \text{by (**)} \\
&= I(x,I(y,y)) && \text{by (EP)} \\
&= I(x,1_L) && \text{by (IP)} \\
&= 1_L && \text{by (RB).}
\end{aligned}
$$
So, by (FNI), $y = 0_L$. Hence, by (**) $I(x,0_L) = 0_L$ and so $N_I(x) = 0_L$. Therefore, by (FNI), $x = 1_L$.
4.
$$
\begin{aligned}
N_I(y) &= I(y,0_L) && \text{by eq. (6)} \\
&= I(y,I(x,y)) && \text{by (**)} \\
&= I(x,I(y,y)) && \text{by (EP)} \\
&= I(x,I(0_L,y)) && \text{by (*)} \\
&= I(0_L,I(x,y)) && \text{by (EP)} \\
&= I(0_L,0_L) && \text{by (**)} \\
&= N_I(0_L) && \text{by eq. (6)} \\
&= 1_L && \text{by (N1).}
\end{aligned}
$$
So, by (FNI), $y = 0_L$. Hence, by (**) $I(x,0_L) = 0_L$ and so $N_I(x) = 0_L$. Therefore, by (FNI), $x = 1_L$.

($\Leftarrow$) If $x = 1_L$ and $y = 0_L$ then by (N1), $I(x,y) = I(1_L,0_L) = N_I(1_L) = 0_L$. $\qquad\square$

4 (S,N) and S-Implications on L

In this section we reproduce, adapted for the bounded lattice framework, the main results on (S,N) and S-implications which can be found in the book [9] and in papers such as [18, 48, 61].

4.1 Basic Definitions and Properties of (S,N) and S-Implications on L

In classical propositional logic, it is sufficient to consider negation and implication, negation and conjunction, or negation and disjunction, because the other connectives can be obtained from any of these pairs. In particular, when we choose as primitive connectives negation, denoted by $\neg$ and disjunction, denoted by $\vee$, the other connectives are found by abbreviation of a logical expression involving these two connectives. An implication $\alpha \to \beta$, is seen as an abbreviation of the formula $\neg \alpha \vee \beta$, for any propositional formulae α and β. This motivated Enric Trillas and Llorenç Valverde in [65] to introduce the class of S-implications, where the implications on the unit interval are obtained from a continuous t-conorm and strong negations. Later, Claudi Alsina and Enric Trillas in [3] proposed the name of (S,N)-implication for the more general case where the t-conorm needs not be continuous, and negations on the unit interval need not be strong. Nowadays, the constraint of the t-conorm to be continuous in S-implications has been abolished. In the following we will adapt these notions and main properties for the context of t-conorms and negations on L.

Definition 13. Let S and N be a t-conorm and a negation on L, respectively. Then the function $I_{S,N} : L \times L \to L$ defined for each $x, y \in L$ by

$$I_{S,N}(x,y) = S(N(x),y) \tag{7}$$

is called a (S,N)-implication on L, and if N is strong then it is called *strong implication* or just *S-implication*. In this case, S and N are said to be the generators of I.

Proposition 20. *Let* $I : L \times L \to L$ *and* ρ *be an L-automorphism. I is an (S,N)-implication on L generated from S and N iff I^{ρ} is an (S,N)-implication on L generated from S^{ρ} and N^{ρ}.*

Proof. $(\Rightarrow)$ Let S and N be the generators of I. Then for any $x, y \in L$,

$$
\begin{aligned}
I^{\rho}(x,y) &= \rho^{-1}(I(\rho(x),\rho(y))) && \text{by eq. (2)} \\
&= \rho^{-1}(S(N(\rho(x)),\rho(y))) && \text{by eq. (7)} \\
&= \rho^{-1}(S(\rho(\rho^{-1}(N(\rho(x)))),\rho(y))) && \\
&= S^{\rho}(N^{\rho}(x),y) && \text{by eq. (2)} \\
&= I_{S^{\rho},N^{\rho}}(x,y).
\end{aligned}
$$

Therefore, by Corollaries 1 and 2, I^{ρ} is an (S,N)-implication on L.

$(\Leftarrow)$ Straightforward from the previous item and because for each $f : L^n \to L$, $(f^{\rho})^{\rho^{-1}} = f$. $\qquad \square$

Corollary 4. *Let* $I : L \times L \to L$ *and* ρ *be an L-automorphism. I is an S-implication on L iff I^{ρ} is an S-implication on L.*

Proof. Straightforward from Propositions 20 and 4. $\qquad \square$

Proposition 21. *Let S and N be a t-conorm and a negation on L, respectively. Then*

1. $I_{S,N}$ is an implication on L.
2. $N_{I_{S,N}} = N$.
3. $I_{S,N}$ satisfies (NP), (EP), (CB) and (RCP) with respect to N.

Proof. We will just prove that $I_{S,N}$ satisfies (CB), the rest is analogous to Proposition 2.4.3. in [9]. For each $x,y \in L$ $y \leq_L N(x) \vee_L y$. So, by Lemma 1, $y \leq_L S(N(x),y) = I_{S,N}(x,y)$. $\qquad\square$

Proposition 22. *Let $I : L \times L \to L$ and N be a negation on L. If I satisfies (FPA) (or (SPI)), (NP), (LCP) for N and (EP) then the function $S_{I,N} : L \times L \to L$ defined for each $x,y \in L$ by*

$$S_{I,N}(x,y) = I(N(x),y) \qquad (8)$$

is a t-conorm.

Proof. For each $x,y,z \in L$,

(S1)   by (LCP), $S_{I,N}(x,y) = I(N(x),y) = I(N(y),x) = S_{I,N}(y,x)$.
(S2)
$$\begin{aligned}
S_{I,N}(S_{I,N}(x,y),z) &= I(N(I(N(x),y)),z) \quad \text{by eq. (8)} \\
&= I(N(z),I(N(x),y)) \quad \text{by (LCP)} \\
&= I(N(x),I(N(z),y)) \quad \text{by (EP)} \\
&= S_{I,N}(x,S_{I,N}(z,y)) \quad \text{by eq. (8)} \\
&= S_{I,N}(x,S_{I,N}(y,z)) \quad \text{by (S1).}
\end{aligned}$$
(S3)   If $x \leq_L y$ then $N(y) \leq_L N(x)$ and so, by (FPA) $I(N(x),z) \leq_L I(N(y),z)$, i.e. $S_{I,N}(x,z) \leq_L S_{I,N}(y,z)$.
(S4)   By (NP), $S_{I,N}(x,0_L) = S_{I,N}(0_L,x) = I(N(0_L),x) = I(1_L,x) = x$. $\qquad\square$

Corollary 5. *Let $I : L \times L \to L$ and N be a strong negation on L. If I satisfies (FPA) (or (SPI)), (NP), (CP) and (EP) then the function $S_{I,N} : L \times L \to L$ defined in equation (8) is a t-conorm.*

Proof. Straightforward from Proposition 22 and Remark 3. $\qquad\square$

Enric Trillas and Llorenç Valverde in Theorem 3.2. [66] (see also Theorem 1.13 in [28] and Theorem 1.6 in [8]) provide a characterization of S-implications on the unit interval which considers properties (NP), (EP) and (CP) and lately, in Theorem 2.6 of [8], the property (CP) was replaced with the property (SNI). We will adapt in the next Theorem, such characterization for S-implications on L.

Theorem 2. *$I : L \times L \to L$ is an S-implication on L iff I satisfies (FPA), (EP) and (SNI).*

Proof. ($\Rightarrow$) Straightforward from Proposition 21.
    ($\Leftarrow$) By (SNI), N_I is a strong negation on L. So, for each $x,y \in L$, (*) $I(x,y) = I(N_I(N_I(x)),y) = S_{I,N_I}(N_I(x),y) = I_{S_{I,N_I},N_I}(x,y)$. On the other hand, because, I satisfies (EP), then by Proposition 19 item 4, I satisfies (CP) for N_I. Therefore, by Proposition 19 item 2, I satisfies (NP). Hence, by Corollary 5, S_{I,N_I} is a t-conorm on L and therefore, from (*), I is an S-implication on L. $\qquad\square$

Remark 4. If S_1 and S_2 are t-conorms on L such that $S_1 \leq_L S_2$ then for any negation N on L, $I_{S_1,N} \leq_L I_{S_2,N}$. Analogously, If N_1 and N_2 are negations on L such that $N_1 \leq_L N_2$ then for any t-conorm S on L, $I_{S,N_1} \leq_L I_{S,N_2}$.

Proposition 23. *Let I be an (S,N)-implication on L. Then $I_{S,N_\perp} \leq_L I \leq_L I_{S,N_\top}$.*

Proof. Note that for any $x, y \in L$ and t-conorm S on L,

$$I_{S,N_\perp}(x,y) = \begin{cases} 1_L & \text{if } x = 0_L, \\ y & \text{otherwise,} \end{cases}$$

$$I_{S,N_\top}(x,y) = \begin{cases} 1_L & \text{if } x \neq 1_L, \\ y & \text{otherwise.} \end{cases}$$

Since I is an (S,N)-implication on L, if $x = 1_L$ then $I(x,y) = S'(N(1_L),y) = y$ and so, clearly $I \leq_L I_{S,N_\top}$. On the other hand, by Proposition 5, $y \leq_L N(x) \vee_L y \leq_L S'(N(x),y) = I(x,y)$ and therefore, clearly, $I_{S,N_\perp} \leq_L I$. □

Notice that for $I_{S,N_\perp}$ and $I_{S,N_\top}$ the t-conorm S has no influence, i.e. for any t-conorms S and S' on L, $I_{S,N_\perp} = I_{S',N_\perp}$ and $I_{S,N_\top} = I_{S',N_\top}$.

4.2 (S,N)-*Implications on L and the Extra Properties*

Notice that, by Proposition 21, all (S,N)-implications on L satisfy (NP), (EP), (CB) and (RCP). But the other properties, i.e. (IP), (OP), (IBL), (CP) and (LCP), can be satisfied by some (S,N)-implications on L and not by other (S,N)-implications on L. In this section we will provide characterizations, under certain conditions, of (S,N)-implications on L which satisfy these properties.

4.2.1 (S,N)-**Implications on L and (IP)**

Not every (S,N)-implication on L satisfies (IP). For example for any bounded lattice L and negation N on L, $I_{GKD}(x,y) = N(x) \vee_L y$ is trivially an (S,N)-implication on L, that, in determined conditions (e.g. N has an equilibrium point) does not satisfy (IP). This (S,N)-implication on L generalizes the well known Kleene-Dienes implication on the unit interval in [40, 25]. An (S,N)-implication on L which satisfies (IP) is $I_{S,N_\top}$. In the following we provide a necessary condition for an (S,N)-implications on L to satisfy (IP).

Proposition 24. *Let L be a complete lattice, S be a t-conorm on L and N be a negation on L. If $I_{S,N}$ satisfies (IP) then $N_S \leq_L N$ and $N_S \circ N \leq_L Id_L$.*

Proof. If $I_{S,N}$ satisfies (IP) then for any $x \in L$, $S(N(x),x) = 1_L$. So, for any $x \in L$, $N(x) \in \{z \in L : S(x,z) = 1_L\}$ and therefore, $N_S(x) = \inf\{z \in L : S(x,z) = 1_L\} \leq_L N(x)$. So, $N_S \leq_L N$.

Since by (IP), $S(N(x),x) = 1_L$ then $N_s(N(x)) = \inf\{z \in L : S(N(x),z) = 1_L\} \leq_L x$. So, $N_S \circ N \leq_L Id_L$. □

The next two propositions provide a characterization, under certain conditions, of (S,N)-implications on L which satisfy (IP).

Proposition 25. *Let L be a complete lattice, S be a positive t-conorm on L and N be a negation on L. $I_{S,N}$ satisfies (IP) iff $N_S = N = N_\top$.*

Proof. ($\Rightarrow$) Straightforward from Lemma 2 and Proposition 24.

($\Leftarrow$) If $x = 1_L$ then trivially $S(N(x),x) = 1_L$. If $x \neq 1_L$, then $N(x) = N_\top(x) = 1_L$ and so $S(N(x),x) = 1_L$. So, $I_{S,N}$ satisfies (IP). $\qquad\qquad\square$

Proposition 26. *Let L be a complete lattice, N be a negation on L and S be a t-conorm on L such that for each $A \subseteq L$, $S(\inf A, y) = \inf\{S(a,y) : a \in A\}$. $I_{S,N}$ satisfies (IP) iff $N_S \leq_L N$.*

Proof. By Proposition 24, we just need to prove the left side of this proposition. But, $S(N_S(x),x) = S(\inf\{z \in L : S(x,z) = 1_L\},x) = \inf\{S(z,x) : S(x,z) = 1_L\} = 1_L$. Since $N_S \leq_L N$ then by (S3), $S(N_S(x),x) \leq_L S(N(x),x)$ and therefore $S(N(x),x) = 1_L$. $\quad\square$

4.2.2 (S,N)-Implications on L and (OP)

Since (OP) implies (IP), I_{GKD} is an example of an (S,N)-implication on L which, under certain conditions, does not satisfy (OP). On the other hand, (IP) does not imply (OP), and then (S,N)-implications on L satisfying (IP) do not necessarily satisfy (OP). In fact, this is the case for $I_{S,N_\top}$. An example of an (S,N)-implication on a bounded lattice which satisfies (OP) is the following: let L be a totally ordered bounded lattice, N be a strong negation on L and $S : L \times L \to L$ defined by

$$S(x,y) = \begin{cases} 1_L & \text{if } N(x) \leq_L y, \\ x \vee_L y & \text{otherwise.} \end{cases}$$

It is not hard to check that S is a t-conorm on L and that $I_{S,N}$ satisfies (OP).

Theorem 3. *Let L be a complete lattice, S be a t-conorm and N be a negation on L. $I_{S,N}$ satisfies (OP) iff I satisfies (IP) and $N_S \circ N = Id_L$.*

Proof. ($\Rightarrow$) (IP) is straightforward from (OP). On the other hand, by (OP) if $S(N(x),z) = 1_L$ then $x \leq_L z$ and therefore $x \leq_L \inf\{z \in L : S(N(x),z) = 1_L\} = N_S(N(x))$. So, by Proposition 24, $N_S \circ N = Id_L$.

($\Leftarrow$) By Proposition 21, I satisfies (SPI). Thus, if $x \leq_L y$, then by (SPI), $I_{S,N}(x,x) \leq_L I_{S,N}(x,y)$ and therefore, by (IP), $I_{S,N}(x,y) = 1_L$. On the other hand, if $I_{S,N}(x,y) = 1_L$ then $S(N(x),y) = 1_L$ and so $N_S(N(x)) \leq_L y$. Therefore, because $N_S \circ N = Id_L$, $x \leq_L y$. $\qquad\qquad\square$

4.2.3 (S,N)-Implications on L and (IBL)

It is clear that not every (S,N)-implication on L satisfies (IBL). For example, for any L and $a \in L - \{0_L, 1_L\}$, $I_{S_\top,N_a}$ where N_a is the negation on L defined by

$$N_a(x) = \begin{cases} a & \text{if } x \neq 0_L \text{ and } x \neq 1_L, \\ 1_L & \text{if } x = 0_L, \\ 0_L & \text{if } x = 1_L, \end{cases}$$

clearly does not satisfy (IBL). In the following, based on the Theorem 10 in [61], we will provide a characterization of the (S,N)-implications on L which satisfy (IBL).

Theorem 4. *Let S be a t-conorm and N be a negation on L. $I_{S,N}$ satisfies (IBL) iff the range of N is a subset of the set of idempotent elements of S.*

Proof. Analogous to the proof of Theorem 10 in [61].   □

Corollary 6. *Let S be a t-conorm and N be a strong negation on L. $I_{S,N}$ satisfies (IBL) iff $S = S_\perp$.*

Proof. Since N is strong its range is L. By Proposition 6, $S_\perp$ is the unique t-conorm on L such that each element in L is idempotent and therefore, by Theorem 4, $I_{S,N}$ satisfies (IBL) iff $S = S_\perp$.   □

4.2.4 (S,N)-Implications on L and (CP)

Although (S,N)-implications on L satisfy (RCP), not all (S,N)-implications on L satisfy (CP) (neither (LCP)). But, since by Remark 3, (CP), (LCP) and (RCP) are equivalent when the negation is strong, then it is clear that a sufficient condition for an (S,N)-implication on L to satisfy (CP) is that the negation be strong. In this subsection we will prove that this requirement is also a necessary condition for an (S,N)-implication to satisfy (CP).

Theorem 5. *Let S be a t-conorm and N be a negation on L. Then $I_{S,N}$ satisfies (CP) for N iff N is strong.*

Proof. ($\Rightarrow$) By Proposition 21, $I_{S,N}$ satisfies (NP) and so for any $x \in L$, $x = I_{S,N}(1_L,x)$. Therefore, by (CP) for N, $x = I_{S,N}(N(x),N(1_L)) = S(N(N(x)),0_L) = N(N(x))$.

($\Leftarrow$) Since N is strong and S commutative, then for each $x,y \in L$, $I(x,y) = I(x,N(N(y))) = S(N(x),N(N(y))) = S(N(N(y)),N(x)) = I(N(y),N(x))$.   □

4.2.5 (S,N)-Implications on L and (LCP)

Analogously to the previous subsection, in this subsection we will prove that a necessary and sufficient condition for an (S,N)-implication on L to satisfy (LCP) is that the negation is strong.

Theorem 6. *Let S be a t-conorm and N be a negation on L. Then $I_{S,N}$ satisfies (LCP) for N iff N is strong.*

Proof. ($\Rightarrow$) By Proposition 21, $I_{S,N}$ satisfies (NP) and so for any $x \in L$, $x = I_{S,N}(1_L,x) = I_{S,N}(N(0_L),x)$. Therefore, by (LCP) for N,

$$x = I_{S,N}(N(x), 0_L) = S(N(N(x)), 0_L) = N(N(x)).$$

($\Leftarrow$) Straightforward from Theorem 5 and Remark 3. □

5 Final Remarks

In this chapter, we have generalized the notion and some properties of implications on the unit interval in the sense of Fodor and Roubens [28], and in particular the class of (S,N)-implications on the unit interval, for implications and (S,N)-implications on bounded lattices. Most of the properties follow analogously to the $[0,1]$-valued case. However some properties, such as Proposition 11, due to the partial order, are just weaker versions of a $[0,1]$-valued property. For other properties, the proof was simplified, as for example in Theorem 3, where we provide a characterization of the (S,N)-implications which satisfy (OP) analogous to Theorem 2.4.19 of [9], but proved in a simpler way. In other cases some conditions were relaxed, such as in Proposition 18. Finally, we also have introduced a new class of negations on L, that of frontier negations on L.

The objective of this paper is to show that several notions and properties of usual fuzzy logics can be naturally extended for a framework where the membership degrees are taken from arbitrary bounded lattices. But it is also to show that in some cases the properties are not completely equivalent to their fuzzy counterpart.

Fuzzy set theory on bounded lattices allows to cover in a single frame, several extensions of the same notions or properties in fuzzy logics. Moreover, each one of the extensions of fuzzy logics (e.g Interval-valued, Atanassov Intuitionistic, fuzzy multisets, etc.), can also be naturally generalized by considering an arbitrary lattice instead of $\langle [0,1], \leq \rangle$. For example, this was done in [15, 23, 55] with the interval-valued extension. We shall do this in a subsequent paper.

Acknowledgements. This work was partially supported by projects CNPq 308256/2009-3 and CNPq 201118/2010-6 (from Government of Brazil), H. Bustine was supported by project TIN 2010-15055 of the Spanish Ministry of Science. A. Pradera was supported by projects TIN2009-07901 and TEC2009-14587-C03-03 from the Spanish Government. The authors wish to thank the anonymous reviewers for their constructive comments which helped to improve the presentation of this chapter.

References

[1] Abbot, J.C.: Semi-Boolean algebras. Mathematichki Vesnik 4, 177–198 (1967)
[2] Alcalde, C., Burusco, A., Fuentes-González, R.: A constructive method for the definition of interval-valued fuzzy implication operators. Fuzzy Sets and Systems 153(2), 211–227 (2005)
[3] Alsina, C., Trillas, E.: When (S,N)-implications are (T,T_1)-conditional functions? Fuzzy Sets and Systems 134(2), 305–310 (2003)
[4] Alsina, C., Trillas, E., Valverde, L.: On non-distributive logical connectives for fuzzy set theory. Busefal 3, 18–29 (1980)

[5]   Atanassov, K.T.: Intuitionistic fuzzy sets. Fuzzy Sets and Systems 20, 87–96 (1986)

[6]   Atanassov, K.T., Gargov, G.: Interval valued intuitionistic fuzzy sets. Fuzzy Sets and Systems 31, 343–349 (1989)

[7]   Baczyński, M.: Residual implications revised. Notes on the Smets-Magrez theorem. Fuzzy Sets and Systems 145(2), 267–277 (2004)

[8]   Baczyński, M., Jayaram, B.: On the characterizations of (S,N)-implications. Fuzzy Sets and Systems 158, 1713–1727 (2007)

[9]   Baczyński, M., Jayaram, B.: Fuzzy Implications. Springer, Heidelberg (2008)

[10] De Baets, B., Mesiar, R.: Triangular norms on product lattices. Fuzzy Sets and Systems 104(1), 61–75 (1999)

[11] Bedregal, B., Beliakov, G., Bustince, H., Calvo, T., Mesiar, R., Paternain, D.: A class of fuzzy multisets with a fixed number of memberships. Information Sciences 189(1), 1–17 (2012)

[12] Bedregal, B.C.: On interval fuzzy negations. Fuzzy Sets and Systems 161, 2290–2313 (2010)

[13] Bedregal, B.C., Dimuro, G.P., Santiago, R.H.N., Reiser, R.H.S.: On interval fuzzy S-implications. Information Sciences 180(8), 1373–1389 (2010)

[14] Bedregal, B.C., Reiser, R.H.S., Dimuro, G.P.: Xor-implications and E-implications: Classes of fuzzy implications based on fuzzy xor. Elec. Notes in Theoretical Computer Science 247, 5–18 (2009)

[15] Bedregal, B.C., Santos, H.S., Callejas-Bedregal, R.: T-norms on bounded lattices: t-norm morphisms and operators. In: Proceedings of 2006 IEEE International Conference on Fuzzy Systems, Vancouver, Canada, pp. 22–28 (2006)

[16] Birkhoff, G.: Lattice Theory. American Mathematical Society, Providence (1973)

[17] Bustince, H., Barrenechea, E., Mohedano, V.: Intuitionistic fuzzy implication operators – an expression and main properties. International Journal of Uncertainty, Fuzziness and Knowledge-Based Systems 12(13), 387–406 (2004)

[18] Bustince, H., Burillo, P., Soria, F.: Automorphisms, negations and implication operators. Fuzzy Sets and Systems 134(2), 209–229 (2003)

[19] Chang, C.C.: Algebraic analisys of many valued logics. Transactions of the American Mathematical Society 88, 467–490 (1958)

[20] De Cooman, G., Kerre, E.E.: Order norms on bounded partially ordered sets. Fuzzy Mathematics 2, 281–310 (1994)

[21] Da Costa, C.G., Bedregal, B.C., Doria Neto, A.D.: Relating De Morgan triples with Atanassov's intuitionistic De Morgan triples via automorphisms. International Journal of Approximate Reasoning 52(4), 473–487 (2010)

[22] Davey, B.A., Priestley, H.A.: Introduction to Lattices and Order, 2nd edn. Cambridge University Press, Cambridge (2002)

[23] Deschrijver, G.: A representation of t-norms in interval-valued L-fuzzy set theory. Fuzzy Sets and Systems 159, 1597–1618 (2008)

[24] Deschrijver, G., Cornelis, C., Kerre, E.E.: On the representation of intuitionistic fuzzy t-norms and t-conorms. IEEE Transactions on Fuzzy Systems 12(1), 45–61 (2004)

[25] Dienes, Z.P.: On an implication function in many-valued systems of logic. Journal of Symbolic Logics 14, 95–97 (1949)

[26] Dujet, C., Vincent, N.: Force implication: A new approach to human reasoning. Fuzzy Sets and Systems 69(1), 53–63 (1995)

[27] Esteva, F., Godo, L.: Monoidal t-norm based logic: Toward a logic for left-continuous t-norms. Fuzzy Sets and Systems 123(3), 271–288 (2001)

[28] Fodor, J., Roubens, M.: Fuzzy Preference Modelling and Multicriteria Decision Support. Kluwer Academic Publisher, Dordrecht (1994)

[29] Font, J.M., Rodriguez, A.J., Torrens, A.: Wajsberg algebra. Stochastica 8(1), 5–31 (1984)

[30] Gödel, K.: Zum intuitionistischen Aussagenkalkül. Anzeiger Akademie der Wissenschaften Wien 69, 65–66 (1932)

[31] Goguen, J.: L-fuzzy sets. Journal of Mathematical Analysis and Applications 18(1), 145–174 (1967)

[32] Grätzer, G.: Lattice Theory: First Concepts and Distributive Lattices. Dover Publications (2009)

[33] Grätzer, G.: Lattice Theory: Foundation. Birkhäuser (2011)

[34] Hájek, P.: Metamathematics of Fuzzy Logic. Springer, Heildelberg (2001)

[35] Hellmann, M.: Fuzzy logic introduction. Epsilon Nought - Radar Remote Sensing Tutorial (2001)

[36] Hungerford, T.W.: Algebra. Springer, New York (1974)

[37] Grattan-Guiness, I.: Fuzzy membership mapped onto interval and many-valued quantities. Z. Math. Logik. Grundladen Math. 22, 149–160 (1975)

[38] Karaçal, F.: On the directed decomposability of strong negations and S-implications operators on product lattices. Information Sciences 176(20), 3011–3025 (2006)

[39] Kitainik, L.: Fuzzy Decision Procedures with Binary Relations. Kluwer Academic Publisher, Dordrecht (1993)

[40] Kleene, S.C.: On a notation for ordinal numbers. Journal of Symbolic Logics 3, 150–155 (1938)

[41] Klement, E.P., Mesiar, R., Pap, E.: Triangular Norms. Kluwer, Dordrecht (2000)

[42] Li, D.C., Li, Y.M., Xie, Y.J.: Robustness of interval-valued fuzzy inference. Information Sciences 181(20), 4754–4764 (2011)

[43] Lin, L., Xia, Z.Q.: Intuitionistic fuzzy implication operators expressions and properties. Journal of Applied Mathematics & Computing 22(3), 325–338 (2006)

[44] Liu, H.-W., Xue, P.-J.: T-Seminorms and Implications on a Complete Lattice. In: Cao, B.-Y., Wang, G.-J., Chen, S.-L., Guo, S.-Z. (eds.) Quantitative Logic and Soft Computing 2010. AISC, vol. 82, pp. 215–225. Springer, Heidelberg (2010)

[45] Łukasiewicz, J.: O logice trójwartościowej. Ruch Filozoficzny 5, 170–171 (1920); English translation: On three-valued logic, In: Borkowski, L. (ed.), Jan Łukasiewicz Selected Works, pp. 87–88. North Holland (1990)

[46] Ma, Z., Wu, W.: Logical operators on complete lattices. Information Sciences 55(1), 77–97 (1991)

[47] Mas, M., Monserrat, M., Torrens, J.: A characterization of (U,N), RU, QL and D-implications derived from uninorms satisfying the law of importation. Fuzzy Sets and Systems 161(10), 1369–1387 (2010)

[48] Mas, M., Monserrat, M., Torrens, J., Trillas, E.: A survey on fuzzy implication functions. IEEE Transactions on Fuzzy Systems 15(6), 1107–1121 (2007)

[49] Massanet, S., Torrens, J.: On a new class of fuzzy implications: h-implications and generalizations. Information Sciences 181(11), 2111–2127 (2011)

[50] Menger, K.: Statistical metrics. Proceedings of the National Academy of Sciences 28(12), 535–537 (1942)

[51] Mesiar, R., Komorníková, M.: Aggregation Functions on Bounded Posets. In: Cornelis, C., Deschrijver, G., Nachtegael, M., Schockaert, S., Shi, Y. (eds.) 35 Years of Fuzzy Set Theory. STUDFUZZ, vol. 261, pp. 3–17. Springer, Heidelberg (2010)

[52] Nguyen, H.T., Walker, E.A.: A First Course in Fuzzy Logics, 2nd edn. CRC Press, Boca Raton (2000)

[53] Palmeira, E.S., Bedregal, B.: Extension of fuzzy logic operators defined on bounded lattices via retractions. Computers and Mathematics with Applications 63(6), 1026–1038 (2011)

[54] Post, E.L.: Introduction to a general theory of elementary propositions. American Journal of Mathematics 43, 163–185 (1921)

[55] Reiser, R.H.S., Dimuro, G.P., Bedregal, B., Santos, H.S., Callejas-Bedregal, R.: S-implications on complete lattice and the interval constructor. Tendências em Matemática Aplicada e Computacional – TEMA 9(1), 143–154 (2008)

[56] Ruan, D., Kerre, E.E.: Fuzzy implication operators and generalized fuzzy method of cases. Fuzzy Sets and Systems 54(1), 23–37 (1993)

[57] Schweizer, B., Sklar, A.: Associative functions and abstract semigroups. Publ. Math. Dedrecen 10, 69–81 (1963)

[58] Shang, Y., Yuan, X., Lee, E.S.: The n-dimensional fuzzy sets and Zadeh fuzzy sets based on the finite valued fuzzy sets. Computers & Mathematics with Applications 60, 442–463 (2010)

[59] Shi, Y., Van Gasse, B., Ruan, D.: Implications in Fuzzy Logic: Properties and a New Class. In: Cornelis, C., Deschrijver, G., Nachtegael, M., Schockaert, S., Shi, Y. (eds.) 35 Years of Fuzzy Set Theory. STUDFUZZ, vol. 261, pp. 83–103. Springer, Heidelberg (2010)

[60] Shi, Y., Van Gasse, B., Ruan, D., Kerre, E.E.: On dependencies and independencies of fuzzy implication axioms. Fuzzy Sets and Systems 161(10), 1388–1405 (2010)

[61] Shi, Y., Ruan, D., Kerre, E.E.: On the characterizations of fuzzy implications satisfying $I(x,y) = I(x,I(x,y))$. Information Sciences 177(14), 2954–2978 (2007)

[62] Sussner, P., Nachtegael, M., Mélange, T., Deschrijver, G., Esmi, E., Kerre, E.E.: Interval-valued and intuitionistic fuzzy mathematical morphologies as special cases of $\mathbb{L}$-fuzzy mathematical morphology. Mathematical Imaging and Vision (2011), doi:10.1007/s10851-011-0283-1

[63] Titani, S.: A lattice-valued set theory. Archive for Mathematical Logic 38, 395–421 (1999)

[64] Trillas, E.: Sobre funciones de negación en la teoria de los conjuntos difusos. Stochastica 3(1), 47–59 (1979)

[65] Trillas, E., Valverde, L.: On some functionally expressable implications for fuzzy set theory. In: Klement, E. (ed.) Proc. Third International Seminar on Fuzzy Set Theory, Linz, Austria, pp. 173–190 (1981)

[66] Trillas, E., Valverde, L.: On Implication and Indistinguishability in the Setting of Fuzzy Logic. In: Management Decision Support Systems Using Fuzzy Sets and Possibility Theory. Verlag TÜV, Rheinland (1985)

[67] Wang, Z., Yu, Y.: Pseudo-t-norms and implications operators on a complete Browerian lattice. Fuzzy Sets and Systems 132, 113–124 (2002)

[68] Xu, Y., Ruan, D., Qim, K., Liu, J.: Lattice-Valued Logic. Springer (2010)

[69] Yager, R.R.: On the implication operator in fuzzy logic. Information Sciences 31(2), 141–164 (1983)

[70] Yager, R.R.: On some new class of implication operators and their role in approximate reasoning. Information Sciences 167, 193–216 (2004)

[71] Zadeh, L.A.: Fuzzy sets. Information and Control 8, 338–353 (1965)

Implication Functions Generated Using Functions of One Variable

Dana Hliněná, Martin Kalina, and Pavol Král'

Abstract. This chapter presents a survey of implication functions generated using an appropriate function of one variable including f-implications and g-implications introduced by Yager, h-generated implications introduced by Jayaram, h-implications and their generalizations introduced by Massanet and Torrens, I_f and I^g implications introduced by Smutná-Hliněná, and Biba.

1 Introduction

A fuzzy implication is a mapping $I : [0,1]^2 \rightarrow [0,1]$ that generalizes the classical implication to fuzzy logic case in a similar way as t-norms (t-conorms) are generalizations of the classical conjunction (disjunction). Using a straightforward generalization of the classical case we get the families of R_T-, (S,N)- and QL-implications ([2]). It is well known that triangular norms can be generated using appropriate mappings of one variable, called generators. This allows us to investigate properties of t-norms, t-conorms etc., studying properties of their generators. Therefore the question whether something similar is possible in the case of fuzzy implications is

Dana Hliněná
Dept. of Mathematics, FEEC Brno Uni. of Technology,
Technická 8, 616 00 Brno, Czech Republic
e-mail: hlinena@feec.vutbr.cz

Martin Kalina
Dept. of Mathematics, Slovak Uni. of Technology,
Radlinského 11, 813 68 Bratislava, Slovakia
e-mail: kalina@math.sk

Pavol Král'
Dept. of Quant. Methods and Inf. Systems, Faculty of Economics,
Matej Bel University, Tajovského 10,
975 90 Banská Bystrica, Slovakia
e-mail: pavol.kral@umb.sk

M. Baczyński et al. (Eds.): *Adv. in Fuzzy Implication Functions*, STUDFUZZ 300, pp. 125–153.
DOI: 10.1007/978-3-642-35677-3_6   © Springer-Verlag Berlin Heidelberg 2013

very interesting from the theoretical point of view. Yager ([20]) introduced two new classes of implications: f-implications and g-implications, where their generators f are continuous additive generators of Archimedean t-norms ([4]) and generators g are continuous additive generators of Archimedean t-conorms ([4]). Massanet and Torrens in [15] presented a similar concept, h-implications generated from an additive generator of a representable uninorm. Jayaram in [11, 12] proposed so-called h-generated implications, where h can be seen as a multiplicative generator of an Archimedean t-conorm. Smutná ([19]) and Biba, Hliněná ([5]) presented an alternative approach where implications are generated using appropriate strictly decreasing or strictly increasing functions and studied basic properties of proposed generated implications. The chapter is organized as follows. In Section 2 we present some basic preliminaries connected to implications. Despite the fact that we assume the reader's familiarity with basic fuzzy logic connectives (t-norms, t-conorms, negations and implications) and related results, we briefly recall their basic definitions, properties and results used in the rest of our paper. In Section 3 we present f- and g-implications introduced by Yager in [20]. Section 4 is devoted to $h-$implications defined by Jayaram in [11, 12]. Sections 5 and 6 are devoted to $h-$implications and their generalizations proposed by Massanet and Torrens in [15]. In section 7 we present I_f- and I^g-implications introduced by Smutná in [19] and later studied by Biba and Hliněná in [5]. We show also some connections between I_f- and I^g-implications on the one hand and (S,N)-implications and R_T-implications on the other hand. Finally, in section 8 we briefly show some connections and differences between the presented types of implication functions.

2   Preliminaries

We recall notations and basic definitions used in the paper. We also briefly mention some important properties and results in order to make this work self-contained. We start with the basic logic connectives.

Definition 1. (see e.g., [8]) A decreasing function $N : [0,1] \to [0,1]$ is called a fuzzy negation if $N(0) = 1, N(1) = 0$. A fuzzy negation N is called

1. strict if it is strictly decreasing and continuous for arbitrary $x \in [0,1]$,
2. strong if it is an involution, i.e., if $N(N(x)) = x$ for all $x \in [0,1]$.

The dual negation based on a fuzzy negation N is given by

$$N^d(x) = 1 - N(1 - x).$$

Some examples of strict and/or strong fuzzy negations are included in the following example. More examples of fuzzy negations can be found in [8].

Example 1. ([3]) The standard negation $N_s(x) = 1 - x$ and the fuzzy negation $N(x) = \sqrt{1 - x^2}$ are strong. The fuzzy negation $N(x) = 1 - x^2$ is strict, but not strong. The Gödel negation N_{G_1} is the least fuzzy negation and dual Gödel negation N_{G_2} is the greatest fuzzy negation, both are non-continuous and hence not strict (nor strong) fuzzy negations:

$$N_{G_1}(x) = \begin{cases} 1 & \text{if } x = 0, \\ 0 & \text{if } x > 0, \end{cases} ; \qquad N_{G_2}(x) = \begin{cases} 1 & \text{if } x < 1, \\ 0 & \text{if } x = 1. \end{cases}$$

Lemma 1. *Let $N : [0,1] \to [0,1]$ be a strict fuzzy negation. Then its dual negation, N^d, is also strict.*

Definition 2. (see, e.g.,[18]) A triangular norm T (t-norm for short) is a commutative, associative, increasing binary operator on the unit interval $[0,1]$, fulfilling the boundary condition $T(x,1) = x$, for all $x \in [0,1]$.

Remark 1. **(a)** T-norms are broadly used as fuzzy connectives which model conjunctions (see e.g., [10, 13]). There are also connectives, called conjunctors, which model not necessarily associative and commutative fuzzy conjunctions. Conjunctors having 1 as neutral element are called also semicopulas.
(b) Note that the dual operator to a t-norm T defined by $S(x,y) = 1 - T(1-x, 1-y)$ is called t-conorm. For more information, see e.g., [13].

In the literature, we can find several definitions of fuzzy implications. In this paper we will use the following one, which is equivalent to the definition introduced by Fodor and Roubens in [8]. For more detail one can consult [2] or [14].

Definition 3. A function $I : [0,1]^2 \to [0,1]$ is called a fuzzy implication if it satisfies the following conditions:

(I1) I is decreasing in its first variable,
(I2) I is increasing in its second variable,
(I3) $I(1,0) = 0, I(0,0) = I(1,1) = 1$.

Remark 2. Since we will deal only with fuzzy connectives in the whole paper, we will skip the word "fuzzy".

Now, we recall definitions of some important properties of implications. The properties of implications are in fact formal properties of two-valued implications, which are transformed to the fuzzy logic case. For more information one can consult [10] or [17]. The last two properties in the following definition, namely (LI) and (WLI), were proposed in [16].

Definition 4. An implication $I : [0,1]^2 \to [0,1]$ satisfies:

(NP) the left neutrality property, or I is called left neutral, if

$$I(1,y) = y; \quad y \in [0,1],$$

(EP) the exchange principle if

$$I(x, I(y,z)) = I(y, I(x,z)) \text{ for all } x, y, z \in [0,1],$$

(IP) the identity principle if

$$I(x,x) = 1; \quad x \in [0,1],$$

(OP) the ordering property if

$$x \leq y \iff I(x,y) = 1; \quad x,y \in [0,1],$$

(CP) the contrapositive symmetry with respect to a given negation N if

$$I(x,y) = I(N(y),N(x)); \quad x,y \in [0,1].$$

(LI) the law of importation with respect to a t-norm T if

$$I(T(x,y),z) = I(x,I(y,z)); \quad x,y,z \in [0,1].$$

(WLI) the weak law of importation with respect to a commutative and increasing function $F : [0,1]^2 \to [0,1]$ if

$$I(F(x,y),z) = I(x,I(y,z)); \quad x,y,z \in [0,1].$$

Definition 5. Let $I : [0,1]^2 \to [0,1]$ be an implication. The function N_I defined by $N_I(x) = I(x,0)$ for all $x \in [0,1]$, is called the natural negation related to I.

(S,N)-implications which are based on t-conorms and negations form one of the well-known classes of implications.

Definition 6. A function $I : [0,1]^2 \to [0,1]$ is called an (S,N)-implication if there exists a t-conorm S and a negation N such that

$$I(x,y) = S(N(x),y), \quad x,y \in [0,1].$$

If N is a strong negation, then I is called a strong implication.

Example 2. We recall well-known implications which are based on the standard negation and basic continuous t-conorms S_M, S_P and S_L.

- $I_{S_M}(x,y) = \max(1-x,y),$ (Kleene – Dienes implication)
- $I_{S_P}(x,y) = 1-x+x.y,$ (Reichenbach implication)
- $I_{S_L}(x,y) = \min(1-x+y,1).$ (Łukasiewicz implication)

All these implications have the same natural negation, the standard negation $N(x) = 1-x$.

The following characterization of (S,N)-implications is from [2].

Theorem 1. (Baczyński and Jayaram [2], Theorem 5.1) *For a function $I : [0,1]^2 \to [0,1]$, the following statements are equivalent:*

- *I is an (S,N)-implication generated from a t-conorm and a continuous (strict, strong) negation N.*
- *I satisfies (I2), (EP) and N_I is a continuous (strict, strong) negation.*

Another way of extending the classical binary implication to the unit interval $[0,1]$ is based on the residuation operator with respect to a left-continuous triangular norm T

$$I_T(x,y) = \max\{z \in [0,1]; T(x,z) \leq y\}.$$

Elements of this class are known as R_T-implications. The characterization of R_T-implications one can find in [8].

Theorem 2. (Fodor and Roubens [8], Theorem 1.14) *For a function $I : [0,1]^2 \to [0,1]$, the following statements are equivalent:*

- *I is an R_T-implication based on a left-continuous t-norm T.*
- *I satisfies (I2), (OP), (EP), and $I(x,.)$ is right-continuous for any $x \in [0,1]$.*

Example 3. For left-continuous t-norms T_M, T_P a T_L we get the following residual implications:

- $I_{T_M}(x,y) = \begin{cases} 1, & \text{if } x \leq y, \\ y, & \text{otherwise,} \end{cases}$ (Gödel implication)
- $I_{T_P}(x,y) = \min\left(\frac{y}{x}, 1\right),$ (Goguen implication)
- $I_{T_L}(x,y) = \min(1 - x + y, 1).$ (Łukasiewicz implication)

We will consider a class of generated implications based on monotone generators. Therefore we will need a pseudo-inverse of such functions.

Definition 7. (see e.g. [13]) Let $f : [0,1] \to [0,\infty]$ be a decreasing function. The function $f^{(-1)}$ which is defined by

$$f^{(-1)}(x) = \sup\{z \in [0,1]; f(z) > x\}$$

is called the pseudo-inverse of f, with the convention $\sup\emptyset = 0$.

Let $g : [0,1] \to [0,\infty]$ be an increasing function. The function $g^{(-1)}$ which is defined by

$$g^{(-1)}(x) = \sup\{z \in [0,1]; g(z) < x\}$$

is called the pseudo-inverse of g, with the convention $\sup\emptyset = 0$.

Lemma 2. ([5]) *Let c be a positive real number. Then the pseudo-inverse of a positive multiple of any monotone function $f : [0,1] \to [0,\infty]$ satisfies*

$$(c \cdot f)^{(-1)}(x) = f^{(-1)}\left(\frac{x}{c}\right).$$

Recently, several possibilities have been proposed on how to generate implications using appropriate one-variable functions. In the following sections we recall the already known classes of generated implications which were proposed in various papers.

Fig. 1 f-implication I_{YG}

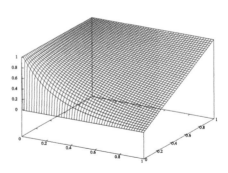

3 Yager's f-Implications and g-Implications

Yager [20] introduced two new families of implications, called f-generated and g-generated implications and discussed their properties as listed in [7] or [8]. Also Jayaram in [11] discussed f-generated implications with respect to three classical logic tautologies, such as distributivity, the law of importation and the contrapositive symmetry.

Proposition 1. ([20]) *If* $f : [0,1] \rightarrow [0,\infty]$ *is a strictly decreasing and continuous function with* $f(1) = 0$, *then the function* $I : [0,1]^2 \rightarrow [0,1]$ *defined by*

$$I(x,y) = f^{-1}(x \cdot f(y)), \quad x,y \in [0,1], \tag{1}$$

with the understanding $0 \cdot \infty = 0$, *is an implication.*

The function f is called an f-generator and the implication represented by (1) is called an f-implication.

Example 4. ([4])

- If we take $f(x) = -\log x$ as the f-generator, which is an additive generator of the product t-norm T_P, then we obtain the Yager implication (Fig. 1):

$$I_{YG}(x,y) = \begin{cases} 1, & \text{if } x = 0 \text{ and } y = 0, \\ y^x, & \text{otherwise,} \end{cases}$$

  which is neither an (S,N)-implication nor an R_T-implication.
- If we take $f(x) = 1 - x$ as the f-generator, which is an additive generator of the Łukasiewicz t-norm $T_L(x,y) = \max(x+y-1,0)$, then we obtain the Reichenbach implication I_{S_P}, which is an (S,N)-implication.

Baczyński and Jayaram in [4] have shown that the generator f-generated implication is obtained from is only unique up to a positive multiplicative constant. They have also investigated the natural negations of the mentioned implications and their relations with (S,N)- and R_T-implications .

Theorem 3. ([4]) *The f-generator of an f-generated implication is uniquely determined up to a positive multiplicative constant, i. e., if f_1 is an f-generator, then f_2 is an f-generator such that $I_{f_1} = I_{f_2}$ if and only if there exists a constant $c \in (0,\infty)$ such that $f_2(x) = c.f_1(x)$ for all $x \in [0,1]$.*

Theorem 4. ([4]) *Let f be an f-generator of an f-generated implication I_f.*

- *If $f(0) = \infty$, then the natural negation N_{I_f} is the Gödel negation N_{G_1}, which is non-continuous.*
- *The natural negation N_{I_f} is a strict negation if and only if $f(0) < \infty$.*
- *I_f is continuous if and only if $f(0) < \infty$.*

Theorem 5. ([20], p. 197) *If f is an f-generator of an f-generated implication I_f , then*

- *I_f satisfies (NP) and (EP),*
- *$I_f(x,x) = 1$ if and only if $x = 0$ or $x = 1$, i. e., I_f does not satisfy (IP),*
- *$I_f(x,y) = 1$ if and only if $x = 0$ or $y = 1$, i. e., I_f does not satisfy (OP),*
- *I_f satisfies (CP) with a negation N if and only if $f(0) < 1$,*
  f_1 defined by $f_1(x) = \frac{f(x)}{f(0)}$, $x \in [0,1]$ is a strong negation and $N = N_{I_f}$.

Theorem 6. ([4]) *If f is an f-generator, then the following statements are equivalent:*

- *I_f is an (S,N)-implication.*
- *$f(0) < \infty$.*

Theorem 7. ([4]) *If f is an f-generator, then I_f is not an R_T-implication.*

Yager [20] has also proposed another class of implications called the g-generated implications. We present their properties in a similar way as in the previous part of this section.

Proposition 2. ([20]) *If $g : [0,1] \rightarrow [0,\infty]$ is a strictly increasing and continuous function with $g(0) = 0$, then the function $I : [0,1]^2 \rightarrow [0,1]$ defined by*

$$I(x,y) = g^{(-1)}\left(\frac{1}{x} \cdot g(y)\right), \quad x,y \in [0,1], \tag{2}$$

with the understanding $\frac{1}{0} = \infty$ and $0 \cdot \infty = \infty$, is an implication.

The function g is called a g-generator and the implication represented by (2) is called a g-implication.

Example 5. ([4])

- If we take the g-generator $g(x) = -\log(1-x)$, which is an additive generator of the probabilistic sum t-conorm S_P, then we obtain the following implication (Fig. 2):

$$I_{YG}(x,y) = \begin{cases} 1, & \text{if } x = 0 \text{ and } y = 0, \\ 1 - (1-y)^{\frac{1}{x}}, & \text{otherwise,} \end{cases}$$

  which is neither an (S,N)-implication nor an R_T-implication.

Fig. 2 g-implication I_{YG}

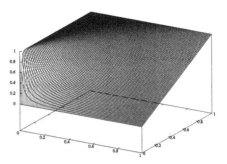

- If we take the g-generator $g(x) = x$, which is a continuous additive generator of the Łukasiewicz t-conorm $S_L(x,y) = \min(x+y,1)$, then we obtain the Goguen implication I_{T_P}, which is an R_T-implication.

Now we present results concerning properties of g-generators, the natural negations of the mentioned implications and their relations with (S,N)- and R_T-implications. More details can be found in [4].

Theorem 8. ([4]) *The g-generator of a g-generated implication is uniquely determined up to a positive multiplicative constant, i. e., if g_1 is a g-generator, then g_2 is a g-generator such that $I_{g_1} = I_{g_2}$ if and only if there exists a constant $c \in (0,\infty)$ such that $g_2(x) = c.g_1(x)$ for all $x \in [0,1]$.*

Theorem 9. ([20], p. 201) *If g is a g-generator of a g-generated implication I_g, then*

- *I_g satisfies (NP) and (EP),*
- *I_g satisfies (IP) if and only if $g(1) < 1$ and $x \le g_1(x)$ for every $x \in [0,1]$, where g_1 is defined by $g_1(x) = \frac{g(x)}{g(1)}$, $x \in [0,1]$,*
- *if $g(1) = 1$, then $I_g(x,y) = 1$ if and only if $x = 0$ or $y = 1$, i. e., I_g does not satisfy (OP) when $g(1) = 1$,*
- *I_g does not satisfy the contrapositive symmetry (CP) with any negation.*

Theorem 10. ([4]) *Let g be a g-generator.*

- *The natural negation of I_g is the Gödel negation N_{G_1}, which is not continuous.*
- *I_g is continuous except at the point $(0,0)$.*

Theorem 11. ([4]) *If g is a g-generator, then I_g is not an (S,N)-implication.*

Theorem 12. ([4]) *If g is a g-generator of I_g, then the following statements are equivalent:*

- *I_g is an R_T-implication.*
- *There exists a constant $c \in]0,\infty[$ such that $g(x) = c \cdot x$ for all $x \in [0,1]$.*
- *I_g is the Goguen R_{T_P}-implication.*

4 Jayaram's *h*-Generated Implications

The f- and g-generators can be seen as the continuous additive generators of t-norms and t-conorms, respectively. A new family of implications called the h-generated implications has been proposed by Jayaram in [12], where h can be seen as a multiplicative generator of a continuous Archimedean t-conorm. In this section we present its definitions, examples and a few of its properties. More details can be found in [4].

Proposition 3 ([12]). *If $h : [0,1] \to [0,1]$ is a strictly decreasing and continuous function with $h(0) = 1$, then the function $I : [0,1]^2 \to [0,1]$ defined by*

$$I(x,y) = h^{(-1)}(x \cdot h(y)), \quad x,y \in [0,1], \tag{3}$$

is an implication.

The function h is called an h-generator and the implication represented by (3) is called an h-generated implication.

Example 6. ([4])

- If we take $h(x) = 1 - x$, which is a continuous multiplicative generator of the probabilistic sum t-conorm S_P, then we obtain the Reichenbach implication $I_{T_{P,}}$, which is an S-implication.
- If we consider the family of h-generators $h_n(x) = 1 - \frac{x^n}{n}, n \in \mathbb{N}$, then we obtain the following implications (Fig. 3):

$$I_n(x,y) = \min\left((n - n \cdot x + x \cdot y^n)^{\frac{1}{n}}, 1\right),$$

which are (S,N)-implications.

Theorem 13. ([4]) *The h-generator of an h-generated implication is uniquely determined, i. e., h_1, h_2 are h-generators such that $I_{h_1} = I_{h_2}$ if and only if $h_1 = h_2$.*

Theorem 14. ([4]) *Let h be an h-generator of I_h.*

Fig. 3 h-generated
implication I_2

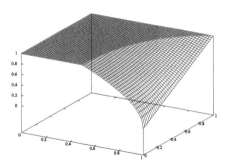

- *The natural negation N_{I_h} is a continuous negation.*
- *I_h is continuous.*

Theorem 15. ([4]) *Let h is an h-generator of an h-generated implication I_h, then*

- *I_h satisfies (NP) and (EP),*
- *I_h satisfies (IP) if and only if $h(1) > 0$ and $x \cdot h(x) \leq h(1)$ for every $x \in [0,1]$,*
- *I_h does not satisfy (OP),*
- *I_h satisfies (CP) with a negation N if and only if $h = h^{-1}$ and $N = N_{I_h}$.*

Theorem 16. ([4]) *If h is an h-generator, then I_h is an (S,N)-implication generated from a t-conorm S and a continuous negation N.*

Theorem 17. ([4]) *If h is an h-generator, then I_h is not an R_T-implication.*

5 Massanet's and Torrens' h- and Generalized h-Implications

In [15] a concept of h-implications was introduced, where h is an additive generator of a representable uninorm. It is a different concept from that presented by B.
Jayaram.

Definition 8. ([15]) *Fix an $e \in]0,1[$ and let $h : [0,1] \to [-\infty,\infty]$ be a strictly increasing and continuous function with $h(0) = -\infty$, $h(e) = 0$ and $h(1) = +\infty$. The function
$I^h : [0,1]^2 \to [0,1]$ defined by*

$$I^h(x,y) = \begin{cases} 1 & \text{if } x = 0, \\ h^{-1}(x \cdot h(y)) & \text{if } x > 0 \text{ and } y \leq e, \\ h^{-1}\left(\frac{1}{x} \cdot h(y)\right) & \text{if } x > 0 \text{ and } y > e, \end{cases} \tag{4}$$

*is called an h-implication. The function h itself is called an h–generator (with respect
to e) of the implication function I^h defined by formula (4).*

Proposition 4. ([15]) *Let h be an h-generator with respect to a fixed $e \in]0,1[$. Then
I^h is an implication.*

Example 7. ([15])

1. Let $h_1(x) = \ln\left(\frac{x}{1-x}\right)$ be an *h*-generator with respect to $\frac{1}{2}$, which is the additive generator of the following conjunctive uninorm

$$U^c_{h_1}(x,y) = \begin{cases} 0, & \text{if } (x,y) \in \{(0,1),(1,0)\}, \\ \frac{xy}{(1-x)(1-y)+xy}, & \text{otherwise.} \end{cases}$$

Then we get the following implication (Fig. 4)

$$I^{h_1}(x,y) = \begin{cases} 1, & \text{if } x = 0, \\ \frac{y^x}{(1-y)^x+y^x}, & \text{if } x > 0 \text{ and } y \le \frac{1}{2}, \\ \frac{y^{\frac{1}{x}}}{(1-y)^{\frac{1}{x}}+y^{\frac{1}{x}}}, & \text{if } x > 0 \text{ and } y > \frac{1}{2}. \end{cases}$$

Fig. 4 I^{h_1} implication

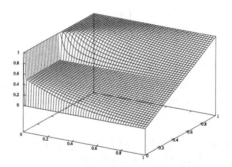

2. Let $h_2(x) = \ln\left(-\frac{1}{\beta}\ln(1-x)\right)$ with $\beta > 0$ be an *h*-generator with respect to $e = 1 - \exp(-\beta)$, of the following disjunctive representable uninorm

$$U^d_{h_2}(x,y) = \begin{cases} 1, & \text{if } (x,y) \in \{(0,1),(1,0)\}, \\ 1 - \exp\left(\frac{1}{\beta}\ln(1-x)\ln(1-y)\right), & \text{otherwise.} \end{cases}$$

Then we get the following implication (Fig. 5)

$$I^{h_2}(x,y) = \begin{cases} 1, & \text{if } x = 0, \\ 1 - \exp\left(-\beta\left(-\frac{1}{\beta}\ln(1-y)\right)^x\right), & \text{if } x > 0 \text{ and } y \le e, \\ 1 - \exp\left(-\beta\left(-\frac{1}{\beta}\ln(1-y)\right)^{\frac{1}{x}}\right), & \text{if } x > 0 \text{ and } y > e. \end{cases}$$

Fig. 5 I^{h_2} implication with $\beta = 2$

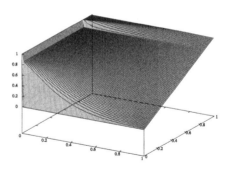

As it is shown in [15], we may relax the conditions $h(0) = -\infty$ and $h(1) = +\infty$.

Definition 9. ([15]) Fix an $e \in]0,1[$ and let $h : [0,1] \to [-\infty,\infty]$ be a strictly increasing and continuous function with $h(e) = 0$ for a fixed $e \in]0,1[$. The function $I^{hg} : [0,1]^2 \to [0,1]$ defined by

$$I^{hg}(x,y) = \begin{cases} 1 & \text{if } x = 0, \\ h^{-1}(x \cdot h(y)) & \text{if } x > 0 \text{ and } y \le e, \\ h^{-1}\left(\min\left\{\frac{1}{x} \cdot h(y), h(1)\right\}\right) & \text{if } x > 0 \text{ and } y > e, \end{cases} \tag{5}$$

is called a generalized h-implication. The function h itself is called a generalized h-generator (with respect to e) of the implication function I^{hg} defined by formula (5).

Proposition 5. ([15]) *Let h be a generalized h-generator with respect to a fixed $e \in]0,1[$. Then I^{hg} is an implication.*

Example 8. ([15])

1. Let $h_1(x) = x - \frac{1}{2}$. Then h_1 is a generalized h-generator satisfying $h_1(0) > -\infty$ and $h_1(1) < \infty$. The generalized h_1-implication is as follows (Fig. 6)

$$I^{h_1g}(x,y) = \begin{cases} 1, & \text{if } x = 0 \text{ or } (x > 0 \text{ and } y \ge \frac{x+1}{2}), \\ x \cdot (y - \frac{1}{2}) + \frac{1}{2}, & \text{if } x > 0 \text{ and } y \le \frac{1}{2}, \\ \frac{1}{x} \cdot (y - \frac{1}{2}) + \frac{1}{2}, & \text{if } x > 0 \text{ and } \frac{1}{2} < y < \frac{x+1}{2}. \end{cases}$$

2. Let $h_2(x) = 1 - \frac{1}{2x}$. Then h_2 is a generalized h-generator satisfying $h_2(0) = -\infty$ and $h_2(1) < \infty$. The generalized h_2-implication is as follows (Fig. 7)

$$I^{h_2g}(x,y) = \begin{cases} 1, & \text{if } x = 0 \text{ or } (x > 0 \text{ and } y \ge \frac{1}{2-x}), \\ \frac{y}{-2xy+x+2y}, & \text{if } x > 0 \text{ and } y \le \frac{1}{2}, \\ \frac{xy}{2xy-2y+1}, & \text{if } x > 0 \text{ and } \frac{1}{2} < y < \frac{1}{2-x}. \end{cases}$$

Fig. 6 I^{h_1g} implication

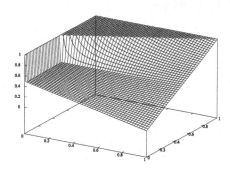

Fig. 7 I^{h_2g} implication

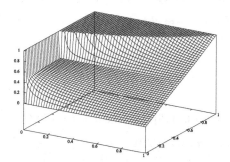

3. Let $h_3(x) = \frac{2x-1}{2-2x}$. Then h_3 is a generalized h-generator satisfying $h_3(0) > -\infty$ and $h(1) = \infty$. The generalized h_3-implication is as follows (Fig. 8)

$$I^{h_3g}(x,y) = \begin{cases} 1, & \text{if } x = 0, \\ \frac{2xy-x-y+1}{2xy-x-2y+2}, & \text{if } x > 0 \text{ and } y \leq \frac{1}{2}, \\ \frac{-xy+x+4y-2}{-2xy+2x+4y-2}, & \text{if } x > 0 \text{ and } y > \frac{1}{2}. \end{cases}$$

Since the behaviour of functions I^h and I^{hg} is very similar, we will discuss only the properties of generalized h-implications. Next lemmas say about the uniqueness of I^{hg} implications and about the natural negations to I^{hg}.

Lemma 3. ([15]) *Let $h_1, h_2; [0,1] \to [-\infty, \infty]$ be two generalized h-generators with respect to a fixed $e \in]0,1[$. Then the following statements are equivalent:*

(a) $I^{h_1,g} = I^{h_2,g}$.

(b) *There exist constants $k, c \in]0, \infty[$ such that*

$$h_2(x) = \begin{cases} k \cdot h_1(x), & \text{if } x \in [0,e[, \\ c \cdot h_1(x), & \text{if } x \in [e,1]. \end{cases}$$

Fig. 8 $I^{h_{3g}}$ implication

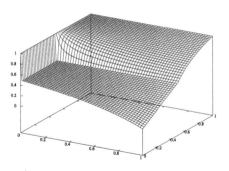

Lemma 4. ([15]) *Let h be a generalized h-generator with respect to a fixed $e \in]0,1[$. Then the following is true for the natural negation $N_{I^{hg}}$ with respect to I^{hg}:*

(a) *If $h(0) = -\infty$, then $N_{I^{hg}} = N_{G_1}$.*
(b) *If $h(0) > -\infty$, then*

$$N_{I^{hg}}(x) = \begin{cases} 1, & \text{if } x = 0, \\ h^{-1}(x \cdot h(0)), & \text{if } x > 0. \end{cases}$$

Theorem 18. ([15]) *Let h be a generalized h-generator.*

  (i) *I^{hg} satisfies (NP).*
 (ii) *$I^{hg}(x,y) \le e \Leftrightarrow (x > 0 \text{ and } y \le e)$.*
(iii) *I^{hg} satisfies (EP).*
(iv) *I^{hg} does not satisfy (OP).*

  - *If $h(1) = \infty$ then $I^{hg}(x,y) = 1$ if and only if $x = 0$ or $y = 1$.*
  - *If $h(1) < \infty$ then $I^{hg}(x,y) = 1$ if and only if $x = 0$ or $y = 1$ or ($e < y < 1$ and $x \cdot h(1) \le h(y)$).*

 (v) *I^{hg} does not satisfy (IP).*

  - *If $h(1) = \infty$ then $I^{hg}(x,x) = 1$ if and only if $x = 0$ or $x = 1$.*
  - *If $h(1) < \infty$ then $I^{hg}(x,x) = 1$ if and only if $x = 0$ or $x = 1$ or ($e < x < 1$ and $x \cdot h(1) \le h(x)$).*

 (vi) *I^{hg} does not satisfy (CP) with any negation.*
(vii) *I^{hg} is continuous except at points $(0,y)$ for $y \le e$.*
(viii) *I^{hg} satisfies (LI) only with respect to T_P.*

Theorem 19. *Let h be a generalized h-generator. Then I^{hg} is neither an (S,N)-implication nor an R_T-implication.*

6 Massanet's and Torrens' (h,e)-Implications and Their Generalization

The motivation behind introducing (h,e)-implications is the property $I(e,y) = y$ for all $y \in [0,1]$, which is a counterpart of (NP) for implications derived from uninorms.

Definition 10. ([15]) Fix an $e \in]0,1[$ and let $h : [0,1] \rightarrow [-\infty,\infty]$ be a strictly increasing and continuous function with $h(0) = -\infty$, $h(e) = 0$ and $h(1) = +\infty$. The function $I^{h,e} : [0,1]^2 \rightarrow [0,1]$ defined by

$$I^{h,e}(x,y) = \begin{cases} 1 & \text{if } x = 0, \\ h^{-1}(\frac{x}{e} \cdot h(y)) & \text{if } x > 0 \text{ and } y \le e, \\ h^{-1}(\frac{e}{x} \cdot h(y)) & \text{if } x > 0 \text{ and } y > e, \end{cases} \tag{6}$$

is called an (h,e)-implication. The function h itself is called an h-generator of the implication function $I^{h,e}$ defined by formula (6).

Proposition 6. ([15]) Let h be an h-generator. Then $I^{h,e}$ defined by (6) is an implication. Moreover, $I^{h,e}(e,y) = y$ for all $y \in [0,1]$.

Proposition 7. ([15]) Let $h_1, h_2; [0,1] \rightarrow [-\infty,\infty]$ be two h-generators with respect to a fixed $e \in]0,1[$. Then the following statements are equivalent:

(a) $I^{h_1,e} = I^{h_2,e}$.
(b) There exist constants $k,c \in]0,\infty[$ such that

$$h_2(x) = \begin{cases} k \cdot h_1(x), & \text{if } x \in [0,e[, \\ c \cdot h_1(x), & \text{if } x \in [e,1]. \end{cases}$$

Example 9. ([15]) Let $h(x) = \ln\left(\frac{x}{1-x}\right)$ be an h-generator with $e = \frac{1}{2}$. Then the (h,e)-implication is (Fig. 9)

$$I^{h,e}(x,y) = \begin{cases} 1, & \text{if } x = 0, \\ \frac{y^{2x}}{(1-y)^{2x}+y^{2x}}, & \text{if } x > 0 \text{ and } y \le \frac{1}{2}, \\ \frac{y^{\frac{1}{2x}}}{(1-y)^{\frac{1}{2x}}+y^{\frac{1}{2x}}}, & \text{if } x > 0 \text{ and } y > \frac{1}{2}. \end{cases}$$

Proposition 8. ([15]) Let h be an h-generator. Then the natural negation with respect to $I^{h,e}$ is the Gödel negation N_{G_1}.

Theorem 20. ([15]) Let h be an h-generator.

(i) $I^{h,e}(x,y) \le e$ if and only if $(x > 0$ and $y \le e)$.
(ii) $I^{h,e}$ satisfies (EP).
(iii) $I^{h,e}$ does not satisfy (IP). $I^{h,e}(x,x) = 1$ if and only if $x = 0$ or $x = 1$.

Fig. 9 $I^{h,e}$ implication

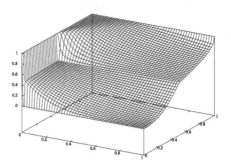

(iv) $I^{h,e}$ *does not satisfy* (OP). $I^{h,e}(x,y) = 1$ *if and only if* $x = 0$ *or* $y = 1$.
(v) $I^{h,e}$ *does not satisfy* (CP) *with respect to any negation.*
(vi) $I^{h,e}$ *is continuous except at points* $(0,y)$ *for* $y \leq e$.
(vii) $I^{h,e}$ *does not satisfy* (WLI) *with respect to any commutative and increasing function* $F : [0,1]^2 \rightarrow [0,1]$. *Consequently* $I^{h,e}$ *does not satisfy* (LI) *with respect to any t-norm.*

Theorem 21. *Let h be an h-generator. Then $I^{h,e}$ is neither an (S,N)-implication nor an R_T-implication.*

Now, we try to relax the conditions $h(0) = -\infty$ and $h(1) = \infty$.

Definition 11. ([15]) *Fix an $e \in]0,1[$ and let $h : [0,1] \rightarrow [-\infty,\infty]$ be a strictly increasing and continuous function with $h(e) = 0$. The function $I^{h_g,e} : [0,1]^2 \rightarrow [0,1]$ defined by*

$$I^{h_g,e}(x,y) = \begin{cases} 1 & \text{if } x = 0, \\ h^{-1}\left(\max\left\{\frac{x}{e} \cdot h(y), h(0)\right\}\right) & \text{if } x > 0 \text{ and } y \leq e, \\ h^{-1}\left(\min\left\{\frac{e}{x} \cdot h(y), h(1)\right\}\right) & \text{if } x > 0 \text{ and } y > e, \end{cases} \tag{7}$$

is called a generalized (h,e)-operator. The function h itself is called an h-generator of the function $I^{h_g,e}$ defined by formula (7).

For $I^{h_g,e}$ we have

$$I^{h_g,e}(1,1) = h^{-1}\left(\min\left\{\frac{e}{1} \cdot h(1), h(1)\right\}\right) = h^{-1}(e \cdot h(1)). \tag{8}$$

This implies that if $h(1) < \infty$ then $I^{h_g,e}$ is not an implication. However, formula (8) does not imply that if $h(0) > -\infty$ then $I^{h_g,e}$ violates conditions for being an implication. At this place Massanet and Torrens made a wrong conclusion that $I^{h_g,e}$ is never an implication.

Proposition 9. *Let h be an h-generator of $I^{h_g,e}$ defined by (7). $I^{h_g,e}$ is an implication if and only if $h(1) = \infty$.*

Implication $I^{h_g,e}$ will be called a *generalized (h,e)-implication.*

Example 10. Let us consider the generalized generator $h_3(x) = \frac{2x-1}{2-2x}$ from Example 8. Then h_3 is a generalized h-generator satisfying $h_3(0) > -\infty$ and $h_3(1) = \infty$ and $e = \frac{1}{2}$. The generalized (h_3, e)-implication is the following (Fig. 10)

$$I^{h_{3g},e}(x,y) = \begin{cases} 1, & \text{if } x = 0, \\ 0, & \text{if } x \geq \frac{1}{2} \text{ and } y \leq \frac{1}{2} - \frac{1}{2(4x-1)}, \\ \frac{4xy-2x-y+1}{4xy-2x-2y+2}, & \text{if } x > 0 \text{ and } \left(\frac{1}{2} - \frac{1}{2(4x-1)}\right) < y \leq \frac{1}{2}, \\ \frac{-xy+x+2y-1}{-2xy+2x+2y-1}, & \text{if } x > 0 \text{ and } y > \frac{1}{2}. \end{cases}$$

Fig. 10 $I^{h_{3g},e}$ implication

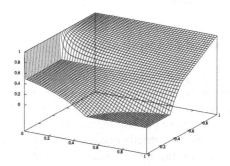

Proposition 10. *Let $h_1, h_2; [0,1] \to [-\infty, \infty]$ be two h-generators with respect to a fixed $e \in]0,1[$ such that $h_1(1) = h_2(1) = \infty$. Then the following statements are equivalent:*

(a) $I^{h_1,g,e} = I^{h_2,g,e}$.
(b) *There exist constants $k, c \in]0, \infty[$ such that*

$$h_2(x) = \begin{cases} k \cdot h_1(x), & \text{if } x \in [0, e[, \\ c \cdot h_1(x), & \text{if } x \in [e, 1]. \end{cases}$$

Proposition 11. *Let h be a generalized h-generator with respect to a fixed $e \in]0,1[$ such that $h(1) = \infty$. Then the following is true for the natural negation $N_{I^{hg},e}$ with respect to $I^{hg,e}$:*

(a) *If $h(0) = -\infty$, then $N_{I^{hg},e} = N_{G_1}$.*
(b) *If $h(0) > -\infty$, then*

$$N_{I^{hg},e}(x) = \begin{cases} 1, & \text{if } x = 0, \\ h^{-1}\left(\max\left\{\frac{x}{e} \cdot h(0), h(0)\right\}\right), & \text{if } x > 0. \end{cases}$$

Theorem 22. *Let h be a generalized h-generator with respect to a fixed $e \in]0,1[$ and with $h(1) = \infty$ and $h(0) > -\infty$.*

(i) $I^{hg,e}$ does not satisfy (EP).

(ii) $I^{hg,e}(x,y) \leq e$ if and only if $x > 0$ and $y \leq e$.

(iii) $I^{hg,e}$ does not satisfy (IP). $I^{hg,e}(x,x) = 1$ if and only if $x = 0$ or $x = 1$.

(iv) $I^{hg,e}$ does not satisfy (OP). $I^{hg,e}(x,y) = 1$ if and only if $x = 0$ or $y = 1$.

(v) $I^{hg,e}$ does not satisfy (CP) with respect to any negation.

(vi) $I^{hg,e}$ is continuous except at points $(0,y)$ for $y \leq e$.

(vii) $I^{hg,e}$ does not satisfy (WLI) with respect to any commutative and increasing function $F : [0,1]^2 \to [0,1]$. Consequently $I^{hg,e}$ does not satisfy (LI) with respect to any t-norm.

Theorem 23. Let h be a generalized h-generator with respect to a fixed $e \in]0,1[$ and with $h(1) = \infty$ and $h(0) > -\infty$. Then $I^{hg,e}$ is neither an (S,N)-implication nor an R_T-implication.

7 I_f^*- and I^g-Implications

Another way of constructing implications is to use residuation operators with respect to Archimedean t-norms. Generators of Archimedean t-norms are strictly decreasing continuous functions $f : [0,1] \to [0,\infty]$ with $f(1) = 0$. However, since implications have just two inputs, we do not need associativity. This means that we may use arbitrary strictly decreasing functions $f : [0,1] \to [0,\infty]$ (with $f(1) = 0$) as generators for the corresponding implication. In a similar way we can generalize (S,N)-implications where S is a generated t-conorm, and use arbitrary strictly increasing function $g : [0,1] \to [0,\infty]$ (with $g(0) = 0$) as generators for the corresponding implication. Smutná in [19] introduced generated implications I_f and I^g. The I_f implications are generated using strictly decreasing functions, the I^g implications are generated using strictly increasing functions. Implications I_f were further discussed in [5]. In the paper we focus on a slightly modified version of I_f implications, which will be denoted by I_f^*. Moreover, we are interested in the connection between I_f^*- and I^g-implications and families of (S,N)-implications and R_T-implications.

7.1   Definitions and Examples

Proposition 12. [5, 19] Let $f : [0,1] \to [0,\infty]$ be a strictly decreasing function such that $f(1) = 0$. Then the function $I_f^*(x,y) : [0,1]^2 \to [0,1]$ which is given by

$$I_f^*(x,y) = f^{(-1)}(\max\{0, f(y) - f(x)\}) \tag{9}$$

is an implication.

Remark 3. (a) The implications I_f ([19]) are defined as

$$I_f(x,y) = \begin{cases} 1 & \text{if } x \leq y, \\ f^{(-1)}(f(y^+) - f(x)) & \text{otherwise,} \end{cases}$$

where the function f has the same meaning as in Proposition 12. We have replaced this formula by (9).

(b) Of course, in the case of a continuous f the corresponding implications I_f^* and I_f coincide. In this case f is an additive generator of an Archimedean t-norm T and I^f is the R_T-implication.

The I_f^*-implications are illustrated by the following examples.

Example 11. ([5], [6]) Let $f_1, f_2, f_3 : [0,1] \to [0,\infty]$ be strictly decreasing functions defined as follows:

- $f_1(x) = \begin{cases} 1-x & \text{if } x \leq 0.5, \\ 0.5-0.5x & \text{otherwise,} \end{cases}$
- $f_2(x) = \frac{1}{x} - 1$,
- $f_3(x) = -\ln(x)$.

Then for $f_1^{(-1)}, f_2^{(-1)}, f_3^{(-1)}$, we get

- $f_1^{(-1)}(x) = \begin{cases} 1-2x & \text{if } x \leq 0.25, \\ 0.5 & \text{if } 0.25 < x \leq 0.5, \\ 1-x & \text{otherwise,} \end{cases}$
- $f_2^{(-1)}(x) = \min\left\{\frac{1}{1+x}, 1\right\}$,
- $f_3^{(-1)}(x) = \min\{e^{-x}, 1\}$,

and the generated implications are (Fig. 11–13)

- $I_{f_1}^*(x,y) = \begin{cases} 1 & \text{if } x \leq y, \\ 1-2x+2y & \text{if } x \leq 0.5, y < 0.5, x-y \leq 0.25, x > y, \\ 0.5 & \text{if } x \leq 0.5, y < 0.5, x-y > 0.25, \\ 0.5 & \text{if } x > 0.5, y < 0.5, x \leq 2y, \\ 0.5+y-0.5x & \text{if } x > 0.5, y < 0.5, x > 2y, \\ 1-x+y & \text{if } x > 0.5, y \geq 0.5, \end{cases}$

- $I_{f_2}^*(x,y) = \begin{cases} 1 & \text{if } x \leq y, \\ \frac{1}{\frac{1}{y}-\frac{1}{x}+1} & \text{otherwise,} \end{cases}$

- $I_{f_3}^*(x,y) = \begin{cases} 1 & \text{if } x \leq y, \\ \frac{y}{x} & \text{otherwise.} \end{cases}$

Implications I^g which are based on strictly increasing functions were introduced in [19].

Fig. 11 $I_{f_1}^*$-implication

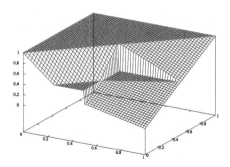

Fig. 12 $I_{f_2}^*$-implication

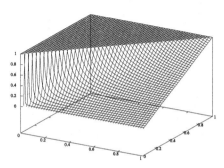

Fig. 13 $I_{f_3}^*$-implication

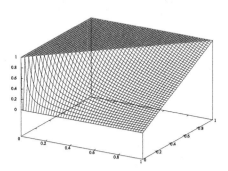

Theorem 24. ([19]) *Let* $g : [0,1] \to [0,\infty]$ *be a strictly increasing function such that* $g(0) = 0$. *Then the function* $I^g(x,y) : [0,1]^2 \to [0,1]$ *which is given by*

$$I^g(x,y) = g^{(-1)}(g(1-x)+g(y)) \tag{10}$$

is an implication.

The above mentioned generated implications I^g are illustrated by the following examples.

Example 12. ([6]) Let $g_1, g_2 : [0,1] \to [0,\infty]$ be given by

- $g_1(x) = \begin{cases} x & \text{if } x \leq 0.5, \\ 0.5 + 0.5x & \text{otherwise,} \end{cases}$
- $g_2(x) = -\ln(1-x)$.

Note that both functions g_1 and g_2 are strictly increasing. For functions $g_1^{(-1)}$ and $g_2^{(-1)}$ we get

- $g_1^{(-1)}(x) = \begin{cases} x & \text{if } x \leq 0,5, \\ 0,5 & \text{if } 0,5 < x \leq 0,75, \\ 2x-1 & \text{if } 0,75 < x \leq 1, \\ 1 & \text{if } 1 < x, \end{cases}$
- $g_2^{(-1)}(x) = 1 - \exp(-x)$ for $x \in [0,\infty]$.

For our functions g_1 and g_2 we have the following (Fig. 14–15)

- $I^{g_1}(x,y) = \begin{cases} 1-x+y & \text{if } x \geq 0.5, y \leq 0.5, x-y \geq 0.5, \\ 0.5 & \text{if } x \geq 0.5, y \leq 0.5, 0.25 \leq x-y < 0.5, \\ 1-2x+2y & \text{if } x \geq 0.5, y \leq 0.5, x-y < 0.25, \\ \min(1-x+2y,1) & \text{if } x < 0.5, y \leq 0.5, \\ \min(2-2x+y,1) & \text{if } x \geq 0.5, y > 0.5, \\ 1 & \text{if } x < 0.5, y > 0.5, \end{cases}$
- $I^{g_2}(x,y) = 1 - \exp(\ln(x(1-y))) = 1 - x + xy$.

Implications I^g may be further generalized. This generalization is based on a replacement of the standard negation by an arbitrary negation.

Theorem 25. ([19]) *Let* $g : [0,1] \to [0,\infty]$ *be a strictly increasing function such that* $g(0) = 0$ *and* N *be a negation. Then* I_N^g, *defined by*

$$I_N^g(x,y) = g^{(-1)}(g(N(x))+g(y)),$$

is an implication.

Fig. 14 I^{g_1}-implication

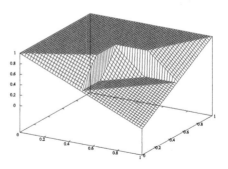

Fig. 15 I^{g_2}-implication

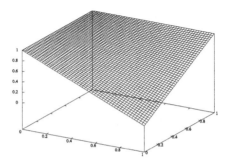

Example 13. ([6]) We deal with strictly increasing functions g_1, g_2, and the negation $N(x) = 1 - x^2$. (Fig. 16–17)

- $I_N^{g_1}(x,y) = \begin{cases} 1 - x^2 + y & \text{if } x \geq \frac{1}{\sqrt{2}}, y \leq 0.5, x^2 - y \geq 0.5, \\ 0.5 & \text{if } x \geq \frac{1}{\sqrt{2}}, y \leq 0.5, 0.25 \leq x^2 - y < 0.5, \\ 1 - 2x^2 + 2y & \text{if } x \geq \frac{1}{\sqrt{2}}, y \leq 0.5, x^2 - y < 0.25, \\ \min(1 - x^2 + 2y, 1) & \text{if } x < \frac{1}{\sqrt{2}}, y \leq 0.5, \\ \min(2 - 2x^2 + y, 1) & \text{if } x \geq \frac{1}{\sqrt{2}}, y > 0.5, \\ 1 & \text{if } x < \frac{1}{\sqrt{2}}, y > 0.5, \end{cases}$

- $I_N^{g_2}(x,y) = 1 - x^2 + x^2 y.$

Obviously, $I_N^{g_1}$ and $I_N^{g_2}$ are implications.

It is well known that generators of continuous Archimedean t-norms and t-conorms are unique up to a positive multiplicative constant, and this is also valid for f and g generators of I_f^*-, I^g- and I_N^g-implications, respectively.

Fig. 16 $I_N^{g_1}$-implication

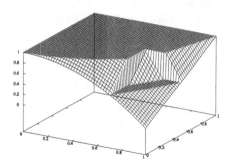

Fig. 17 $I_N^{g_2}$-implication

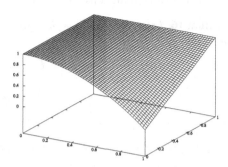

Proposition 13. *Let c be a positive constant.*

(a) *If $f : [0,1] \to [0,\infty]$ is a strictly decreasing function with $f(1) = 0$, then the implications I_f^* and $I_{c \cdot f}^*$ coincide.*

(b) *Assume that $g : [0,1] \to [0,\infty]$ is a strictly increasing function with $g(0) = 0$ and N is an arbitrary negation. Then the implication I^g coincides with $I^{c \cdot g}$, and the implication I_N^g coincides with $I_N^{c \cdot g}$.*

Because of this fact, if f and/or g are bounded functions we can always assume that $f(0) = 1$ and/or $g(1) = 1$, respectively.

7.2 Properties of I_f^*-Implications

In this section we further investigate the properties of I_f^*-implications. We turn our attention to relations between (S,N)-implications and R_T-implications on the one hand and I_f^*-implications on the other hand. By Definition 4 and the following equivalence for a strictly decreasing function f

$$f^{(-1)}(x_0) = 1 \iff x_0 \leq \lim_{x \to 1^-} f(x) = f(1^-) \tag{11}$$

we get directly a condition under which I_f^* satisfies (NP).

Proposition 14. ([5]) *Let* $f : [0,1] \to [0,\infty]$ *be a strictly decreasing function such that* $f(1) = 0$. *Then* I_f^* *satisfies (IP) and (NP). Moreover,* f *is continuous at* $x = 1$ *if and only if* I_f^* *satisfies (OP).*

The continuity of a strictly decreasing function f implies that $f\left(f^{(-1)}(x)\right) = x$. Therefore we can formulate the following proposition. The implication "⇒" was proved in [5].

Proposition 15. *Let* $f : [0,1] \to [0,\infty]$ *be a strictly decreasing function continuous at 1 and 0, such that* $f(1) = 0$. *Then the implication* I_f^* *satisfies (EP) if and only if* f *is continuous.*

In Proposition 15 we have considered just the case when f is continuous at 0 and 1. Now, we will deal with fuctions f being non-continuous at 0 and/or at 1.

Proposition 16. *Let* $f : [0,1] \to [0,\infty]$ *be a strictly decreasing function with* $f(1) = 0$ *discontinuous at 1. Then*

(**a**) *If* $2f(1^-) \geq f(0)$, *then* I_f^* *satisfies (EP). Moreover,*

$$I_f^*(x,y) = \begin{cases} y, & \text{if } x = 1, \\ 1, & \text{otherwise.} \end{cases} \tag{12}$$

(**b**) *If* $2f(1^-) \geq f(0^+)$ *and* $2f(0^+) \leq f(0)$, *then* I_f^* *satisfies (EP). Moreover,*

$$I_f^*(x,y) = \begin{cases} y, & \text{if } x = 1, \\ 0, & \text{if } x \neq 0 \text{ and } y = 0, \\ 1, & \text{otherwise.} \end{cases} \tag{13}$$

Proposition 17. *Let* $f : [0,1] \to [0,\infty]$ *be a strictly decreasing function discontinuous at 0 and continuous at 1 with* $f(1) = 0$. *Then* I_f^* *satisfies (EP) if and only if* f *is continuous in* $]0,1]$ *and fulfils the inequality* $2f(0^+) \leq f(0)$. *Moreover, in this case* $N_{I_f^*} = N_{G_1}$.

Example 14. Define $f : [0,1] \to [0,\infty[$ by the following

$$f(x) = \begin{cases} 1 - x^2, & \text{if } x \in]0,1], \\ 1.5, & \text{if } x = 0. \end{cases}$$

Then $f^{(-1)} : [0,\infty[\to [0,1]$ is given by

$$f^{(-1)}(x) = \begin{cases} \sqrt{1-x}, & \text{if } x \in [0,1[, \\ 0, & \text{if } x \geq 1. \end{cases}$$

Let us compute $I_f^*(0.7, I_f^*(0.8,0))$ and $I_f^*(0.8, I_f^*(0.7,0))$.

$$I_f^*(0.8,0) = f^{(-1)}(1.5 - (1 - 0.8^2)) = 0,$$
$$I_f^*(0.7,0) = f^{(-1)}(1.5 - \sqrt{1 - 0.7^2}) = f^{(-1)}(0.99) = 0.1,$$

and

$$I_f^*(0.7, I_f^*(0.8,0)) = I_f^*(0.7,0) = 0.1,$$
$$I_f^*(0.8, I_f^*(0.7,0)) = I_f^*(0.8,0.1) = \sqrt{0.63} \doteq 0.79,$$

i.e (EP) is violated for I_f^*.

We study the properties of implications I_f^*, under which they are (S,N)- or R_T-implications. Because there are relations between (S,N)- implications and (EP) and the continuity of $N_{I_f^*}$, Propositions 15, 16 and 17 lead us to dealing with continuous function f. The continuity of a generator f implies the continuity of the corresponding natural negation based on I_f^*. Moreover, for a continuous and bounded strictly decreasing function f such that $f(1) = 0$ and $f(0) = c$ the natural negation $N_{I_f^*}$ is strong.

Proposition 18 ([5]). *Let $f : [0,1] \to [0,c]$ be a continuous bounded strictly decreasing function such that $f(1) = 0$. The I_f^* possesses (CP) only with respect to its natural negation $N_{I_f^*}(x) = f^{-1}(f(0) - f(x))$.*

The continuity of a generator f implies that I_f^* is an R_T-implication ([8], Theorem 1.16).

Corollary 1. *Let $f : [0,1] \to [0,c]$ be a strictly decreasing continuous and bounded function with $f(1) = 0$. Then $I_f^*(x,y) = I_f^*(I_f^*(y,0), I_f^*(x,0))$.*

A strictly decreasing continuous function f can be used as an additive generator of a t-norm T and as a generator of an implication I_f^* at the same time. Therefore the relation between the t-norm T and the implication I_f^*, generated by the same function f, is interesting.

Proposition 19. *Let $f : [0,1] \to [0,\infty]$ be a strictly decreasing continuous function such that $f(1) = 0$. If f is an additive generator of a t-norm T, then I_f^* satisfies (LI) with respect to T.*

Proposition 20. *Let $f : [0,1] \to [0,\infty]$ be a strictly decreasing continuous function and $f(1) = 0$. Let T be a t-norm such that $T(x,y) \le f^{(-1)}(f(x) + f(y))$, then the following inequalities hold:*

1, $I_f^*(x, I_f^*(y,z)) \le I_f^*(T(x,y),z)$,
2, $T(I_f^*(x,z), I_f^*(y,z)) \le I_f^*(T(x,y),z)$,
3, $T(I_f^*(x,y), I_f^*(y,z)) \le I_f^*(x,z)$,
4, $T(x, I_f^*(x,y)) \le y$.

Proposition 21. *Let $f : [0,1] \to [0,\infty]$ be a continuous strictly decreasing function such that $f(1) = 0$. Then the implication I_f^* is continuous.*

Let $f : [0,1] \to [0,\infty]$ and $\varphi : [0,1] \to [0,1]$ be arbitrary functions. We will denote

$$(f \circ \varphi)(x) = f(\varphi(x)) \quad \text{for all } x \in [0,1].$$

Proposition 22. *Let $f : [0,1] \to [0,\infty]$ be a strictly decreasing function such that $f(1) = 0$. Let $\varphi : [0,1] \to [0,1]$ be a strictly increasing automorphism. Then the function $(I_f^*)_\varphi(x,y) = \varphi^{-1}(I_f^*(\varphi(x),\varphi(y)))$ is an implication $I_{f \circ \varphi}^*(x,y)$.*

Theorem 26. ([5]) *Let $f : [0,1] \to [0,\infty]$ be a continuous strictly decreasing function such that $f(1) = 0$. Then I_f^* is an R_T-implication given by a continuous t-norm. Moreover, if $f(0) < \infty$, then I_f^* is an (S,N)-implication.*

7.3   Properties of I^g-Implications

In this section we investigate the relations between I^g-implications and (S,N)-implications. Moreover, we study the relations between I^g- and I_f-implications. The following lemma and next two propositions are implied by Theorem 24 and by the fact that $g^{(-1)}(g(x)) = x$ for a strictly monotonous function g.

Lemma 5. *Let $g : [0,1] \to [0,\infty]$ be a strictly increasing function such that $g(0) = 0$. Then the natural negation related to I_g is $N_{I^g}(x) = 1 - x$.*

Proposition 23. *Let $g : [0,1] \to [0,\infty]$ be a strictly increasing function such that $g(0) = 0$. Then I^g satisfies (NP) and (CP) with respect to N_s.*

Proposition 24. *Let $g : [0,1] \to [0,\infty]$ be a continuous and strictly increasing function such that $g(0) = 0$. Then I^g satisfies (EP).*

There exist non-continuous functions g such that I^g satisfy (EP). It is illustrated by the following example.

Example 15. Let $g : [0,1] \to [0,1]$ be given by

$$g(x) = \begin{cases} 0 & x = 0, \\ \frac{1}{2}(x+1) & \text{otherwise.} \end{cases}$$

For its pseudo-inverse we get

$$g^{(-1)}(x) = \begin{cases} 0 & x \leq \frac{1}{2}, \\ 2x - 1 & x > \frac{1}{2}. \end{cases}$$

Because g is continuous in $]0,1]$ and strictly monotone, we have to show that (EP) holds for I^g only for triples (x,y,z) such that $x = 1$ or $y = 1$ or $z = 0$, and it follows from strict monotonicity of g and equality $g^{(-1)} \circ g(x) = x$.

From Theorem 1 and the previous propositions we get the following relation between I^g implications and (S,N)-implications:

Theorem 27. *Let* $g : [0,1] \to [0,\infty]$ *be a strictly increasing function continuous on* $]0,1]$ *such that* $g(0) = 0$. *Then* I^g *is an* (S,N)-*implication which is strong.*

A relation between I_f and I^g is stated by the following proposition.

Proposition 25. ([19]) *Let* $f : [0,1] \to [0,\infty]$ *be a left-continuous strictly decreasing function such that* $f(1) = 0$, *and* $g : [0,1] \to [0,\infty]$ *be a right-continuous strictly increasing function such that* $g(0) = 0$. *Then*

$$I^g(x,y) = I_f(x,y) = \sup\{z \in [0,1]; C(x,z) \le y\}, \tag{14}$$

where $C(x,z) = f^{(-1)}(f(x) + f(z)))$, *and* $g(x) = f(1-x)$.

Remark 4. Implications I_f^* are residual with respect to a left-continuous conjunctor C only if the corresponding generator f is continuous in $[0,1]$.

8 Concluding Notes

Implications generated by functions of one variable have become a rapidly developing branch of fuzzy logics. Our main aim was to present a partial overview of possibilities how to generate implications using additive generators of t-norms, t-conorms and representable uninorms. Since implications are just binary operations, their associativity is meaningless. In this context we presented also a possibility of generating implications using strictly monotone (not necessarily continuous) functions with a range in $[0,\infty]$. Properties of generated families are listed in Table 1. Connections between generated families and (S,N)- and/or R_T-implications are listed in Table 2.

Table 1 Properties of implications generated by one-variable functions

Family	Properties
Yager's f-implications	(NP), (EP)
Yager's g-implications	(NP), (EP)
Jayaram's h-implications	(NP), (EP)
Massanet's and Torrens' h_g-implications	(NP), (EP)
Massanet's and Torrens' (h,e)-implications	(EP)
Massanet's and Torrens' (h_g,e)-implications	—
I_f^*-implications	(NP), (IP)
I^g-implications	(NP), (CP)

Table 2 Implications generated by one-variable functions and (S,N)- and R_T-implications

Family	(S,N)-implication	R_T-implication
Yager's f-implications	if $f(0) < \infty$	—
Yager's g-implications	—	if $g(x) = c \cdot x$ for a $c \in]0, \infty[$
Jayaram's $h-$implications	always	—
Massanet's and Torrens' h_g-implications	—	—
Massanet's and Torrens' (h,e)-implications	—	—
Massanet's and Torrens' (h_g, e)-implications	—	—
I_f^*-implications	—	if f is continuous
I^g-implications	if g is continuous	—

Acknowledgements. Dana Hliněná has been supported by Project MSM0021630529 of the Ministry of Education and project FEKT-S-11-2(921). Martin Kalina has been supported from the Science and Technology Assistance Agency under contract No. APVV-0073-10, and from the VEGA grant agency, grant numbers 1/0143/11 and 1/0297/11. Pavol Král' acknowledges the support from the grant 1/0297/11 provided by the VEGA grant agency.

References

1. Baczyński, M., Jayaram, B.: QL-implications: Some properties and intersections. Fuzzy Sets and Systems 161, 158–188 (2010)
2. Baczyński, M., Jayaram, B.: Fuzzy implications. STUDFUZZ, vol. 231. Springer, Berlin (2008)
3. Baczyński, M., Jayaram, B.: (S,N)- and R-implications: A state-of-the-art survey. Fuzzy Sets and Systems 159, 1836–1859 (2008)
4. Baczyński, M., Jayaram, B.: Yager's classes of fuzzy implications: some properties and intersections. Kybernetika 43, 157–182 (2007)
5. Biba, V., Hliněná, D.: Generated fuzzy implications and known classes of implications. Acta Univ. M. Belii, Ser. Math. 16, 25–34 (2010)
6. Biba, V., Hliněná, D.: Intersection of generated (S,N)-implications and R-implications. In: XXVIIIth International Colloquium on the Management of Educational Process, Brno, pp. 1–10 (2010)
7. Dubois, D., Prade, H.: Fuzzy sets in approximate reasoning. Part I. Inference with possibility distributions. Fuzzy Sets and Systems 40, 143–202 (1991)
8. Fodor, J., Roubens, M.: Fuzzy preference modelling and multicriteria decision support. Kluwer Academic Publishers (1994)
9. Grzegorzewski, P.: On the Properties of Probabilistic Implications. In: Melo-Pinto, P., Couto, P., Serôdio, C., Fodor, J., De Baets, B. (eds.) Eurofuse 2011. AISC, vol. 107, pp. 67–78. Springer, Heidelberg (2011)
10. Hájek, P.: Mathematics of Fuzzy Logic. Kluwer, Dordrecht (1998)
11. Jayaram, B.: Yager's new class of implications and some classical tautologies. Information Sciences 177, 930–946 (2007)
12. Jayaram, B.: Contrapositive symmetrisation of fuzzy implications–revisited. Fuzzy Sets and Systems 157, 2291–2310 (2006)
13. Klement, E.P., Mesiar, R., Pap, E.: Triangular Norms, 1st edn. Springer (2000)

14. Mas, M., Monserrat, J., Torrens, M., Trillas, E.: A survey on fuzzy implication functions. IEEE T. Fuzzy Systems 15, 1107–1121 (2007)
15. Massanet, S., Torrens, J.: On a new class of fuzzy implications: h-implications and generalizations. Information Sciences 181, 2111–2127 (2011)
16. Massanet, S., Torrens, J.: The law of importation versus the exchange principle on fuzzy implications. Fuzzy Sets and Systems 168(1), 47–69 (2011)
17. Novák, V., Perfilieva, I., Močkoř, J.: Mathematical Principles of Fuzzy Logic. Kluwer, Boston (1999)
18. Schweizer, B., Sklar, A.: Probabilistic Metric Spaces. North Holland, New York (1983)
19. Smutná, D.: On many valued conjunctions and implications. Journal of Electrical Engineering 50, 8–10 (1999)
20. Yager, R.R.: On some new classes of implication operators and their role in approximate reasoning. Information Sciences 167, 193–216 (2004)

Compositions of Fuzzy Implications

Józef Drewniak and Jolanta Sobera

Abstract. This chapter considers fuzzy implications as a special case of fuzzy relations in $[0, 1]$. We examine compositions of fuzzy implications based on a binary operation $*$ and discuss the dependencies between algebraic properties of the operation $*$ and the induced $\sup -*$ composition. Under some simple assumptions the $\sup -*$ composition of fuzzy implications gives also a fuzzy implication. This leads to an examination of ordered groupoids and ordered semigroups of fuzzy implications. Contrapositive and invariant fuzzy implications are also considered.

1 Introduction

Fuzzy implications take attention from the beginning of fuzzy sets theory. Their applicational sources are connected with approximate reasoning [22] and fuzzy control [20]. Simultaneously, every fuzzy implication function is an example of the logical connective of multivalued logic [19]. The simplest axiomatization of fuzzy implications was presented by Fodor and Roubens [14] pp. 21-31 (cf. also Gottwald [17]), and its examinations are summarized by Baczyński and Jayaram [5]. A consideration of fuzzy implications as a special case of fuzzy relations in $[0, 1]$ was introduced by Baldwin and Pilsworth [6]. They generated new fuzzy implications as sup-min composition of given ones. Their approach was applied in [3], [4] and [12]. We are going to continue this way for $\sup -*$ composition under additional assumptions about the operation $* : [0, 1]^2 \to [0, 1]$. At first, we recall auxiliary

Józef Drewniak
Institute of Mathematics, University of Rzeszów,
35–310 Rzeszów, ul. Rejtana 16a, Poland
e-mail: jdrewnia@univ.rzeszow.pl

Jolanta Sobera
Institute of Mathematics, University of Silesia,
40-007 Katowice, ul. Bankowa 14, Poland
e-mail: jolanta.sobera@us.edu.pl

M. Baczyński et al. (Eds.): *Adv. in Fuzzy Implication Functions*, STUDFUZZ 300, pp. 155–176.
DOI: 10.1007/978-3-642-35677-3_7   © Springer-Verlag Berlin Heidelberg 2013

properties of binary operations and binary fuzzy relations (Sections 2, 3). Next, the simplest axiomatic definition of fuzzy implications with suitable examples is described (Section 4). Finally, the $\sup-*$ composition of fuzzy implications is examined in the family of all fuzzy implications and in some important subfamilies (Sections 5-7). The main attention is paid to a description of ordered algebraic structures of fuzzy implications.

2 Binary Operations in $[0, 1]$

We recall here some important properties and examples of binary operations in the unit interval. Their examination is strictly connected with diverse applications in fuzzy set theory. Because of their algebraic and functional properties we shall use mixed notation. In many algebraic considerations we use a grupoid $([0,1], *)$ with the internal operation $*$. Simultaneously, many examples of such operations have traditional description as functions $F : [0, 1]^2 \to [0, 1]$ (e.g. triangular norms and fuzzy implications). Monotonicity properties (increasing, decreasing) are used with weak inequalities $(\leqslant, \geqslant)$. The operation $*$ is idempotent if $x * x = x$ for $x \in [0, 1]$.

Example 1. The most natural increasing operation in $[0, 1]$ is the restriction of real numbers product (denoted as the function T_P). Moreover, every mean restricted to $[0, 1]$ is an internal operation in $[0, 1]$, e.g.

$$\min(x, y) = \min\{x, y\}, \ \max(x, y) = \max\{x, y\} \quad \text{(lattice operations)},$$
$$P_1(x, y) = x, \ P_2(x, y) = y \quad \text{(binary projections)},$$
$$A(x, y) = 0.5(x + y), \ G(x, y) = \sqrt{xy} \quad \text{(arithmetic and geometric mean)}$$

for $x, y \in [0, 1]$. The above six operations are idempotent, while T_P has only two idempotent points: $T_P(0, 0) = 0$, $T_P(1, 1) = 1$.

Lemma 1 ([10], Lemma 5). *An operation* $* : [0, 1]^2 \to [0, 1]$ *is increasing if and only if it is distributive with respect to* max. *In particular, the following conditions are equivalent*

$$\underset{a,b,c\in[0,1]}{\forall} (b \leqslant c) \Rightarrow (a * b \leqslant a * c, \ b * a \leqslant c * a), \tag{1}$$

$$\underset{a,b,c\in[0,1]}{\forall} a * \max(b, c) = \max(a * b, a * c), \ \max(b, c) * a = \max(b * a, c * a). \tag{2}$$

Lemma 2 ([9], Lemmas 1, 7). *Let an operation* $*$ *be increasing (cf. (1)). We have*

$$(* \leqslant \min) \Leftrightarrow (\underset{a\in[0,1]}{\forall} a * 1 \leqslant a, \ 1 * a \leqslant a) \Rightarrow (\underset{a\in[0,1]}{\forall} a * 0 = 0 * a = 0). \tag{3}$$

The existence of the neutral element $e = 1$ *implies the existence of the zero element* $z = 0$. *The operation* $*$ *has the zero element* $z = 0$ *if and only if* $0 * 1 = 1 * 0 = 0$.

Definition 1. Let L be a complete lattice with an additional binary operation $*$. The operation $*$ is infinitely distributive with respect to supremum (infinitely sup-distributive) if it fulfils

$$\underset{a,b_t \in L}{\forall}\ a * (\sup_{t \in T} b_t) = \sup_{t \in T}(a * b_t), \qquad \underset{a,b_t \in L}{\forall}\ (\sup_{t \in T} b_t) * a = \sup_{t \in T}(b_t * a) \qquad (4)$$

for an arbitrary index set $T \neq \emptyset$. In particular, the lattice L is called infinitely sup-distributive if the operation $* = \wedge$ is infinitely sup-distributive and the infinite inf-distributivity is defined dually. The lattice L is called infinitely distributive if it is infinitely sup $-$ and inf-distributive.

If card $T = 2$, then (4) reduces to (2). By analogy to Lemma 1, we get

Corollary 1. *Let L be a complete lattice with additional binary operation $*$. If the operation $*$ is infinitely* sup-*distributive, then it is distributive with respect to* max *and it is increasing.*

According to [18] (Proposition 1.22), in the case $L = [0,1]$ we have

Lemma 3. *An operation $* : [0,1]^2 \to [0,1]$ is infinitely* sup-*distributive if and only if it is increasing and left-continuous.*

Definition 2 ([18], pp. 4, 222; cf. [13], Definition 6.4). An increasing binary operation $*$ in $[0,1]$ is called
• a semicopula (triangular seminorm) if it has the neutral element $e = 1$;
• a pseudo triangular norm if it is associative semicopula;
• a triangular norm if it is associative, commutative with neutral element $e = 1$;
• a conjunctive uninorm if it is associative, commutative with neutral element $e \in (0,1]$ and $0 * 1 = 1 * 0 = 0$.

Example 2. The greatest semicopula (pseudo triangular norm, triangular norm) is given by $T_M = \min$. The least semicopula (pseudo triangular norm, triangular norm) is called drastic triangular norm,

$$T_D(x,y) = \begin{cases} \min(x,y), & \text{if } \max(x,y) = 1 \\ 0, & \text{otherwise} \end{cases}, \ x,y \in [0,1]. \qquad (5)$$

We have $T_D \leqslant T_L \leqslant T_P \leqslant T_M$, where $T_L(x,y) = \max(0, x+y-1)$, (Łukasiewicz triangular norm), and $T_P(x,y) = xy$ for $x,y \in [0,1]$.

As examples of conjunctive uninorms for fixed $e \in (0,1)$ we consider the least uninorm $\underline{U}_e$ and the least idempotent uninorm $U_e^{\min}$ (cf. [15]), where $(x,y \in [0,1])$

$$\underline{U}_e(x,y) = \begin{cases} 0 & \text{if } x,y \in [0,e) \\ \max(x,y), & \text{if } x,y \in [e,1] \\ \min(x,y), & \text{otherwise} \end{cases}, \ U_e^{\min}(x,y) = \begin{cases} \max(x,y), & \text{if } x,y \in [e,1] \\ \min(x,y), & \text{otherwise} \end{cases}.$$

For the boundary case $e = 1$ we obtain $\underline{U}_1 = T_D$ and $U_1^{\min} = T_M$.

3   Relation Compositions

Fuzzy relations generalize characteristic functions of binary relations. Let $X, Y \neq \emptyset$ and $L = (L, \vee, \wedge, 0, 1)$ be a complete lattice. An L−fuzzy relation between sets X and Y is an arbitrary mapping $R : X \times Y \to L$ (a fuzzy relation for $L = [0, 1]$). In the case $X = Y$ we say about L−fuzzy relation on a set X. The family of all L−fuzzy relations on X is denoted by $LR(X)$ ($FR(X)$ for fuzzy relations). We are interested in L−fuzzy relations on X for $X = L$. For $R, S \in LR(X)$ we use the induced order and the lattice operations:

$$R \leq S \quad \Leftrightarrow \quad \underset{x,y \in X}{\forall} \ (R(x, y) \leq S(x, y)), \tag{6}$$

$$(R \vee S)(x, y) = R(x, y) \vee S(x, y), \quad (R \wedge S)(x, y) = R(x, y) \wedge S(x, y), \ x, y \in X. \tag{7}$$

Usually these operations are considered as the simplest version of inclusion, sum and intersection of fuzzy relations, respectively (cf. [21]). However, the most important operation on fuzzy relations is their composition.

Definition 3 ([16]). Let L be a complete lattice and $* : L^2 \to L$. By $\sup -*$ composition of L−fuzzy relations R, S we call the L−fuzzy relation $R \circ S$, where

$$(R \circ S)(x, z) = \sup_{y \in X}(R(x, y) * S(y, z)), \quad x, z \in X. \tag{8}$$

In order to distinguish the case $* = \min$ we shall write $R \bullet S$.

Directly from properties of supremum we get

Corollary 2. *The composition* (8) *is increasing with respect to the operation* $*$, *i.e. if* $* \leqslant *'$, *then* $\circ \leqslant \circ'$. *In particular, if* $* \leqslant \min$, *then* $\circ \leqslant \bullet$ (*cf.* (3)).

Properties of composition $\circ$ depends on properties of operation $*$, what was examined in details in the paper [10] in the case $L = [0, 1]$. We recall here some of these results.

Definition 4. Let $c \in [0, 1]$. Fuzzy relation $T_c(x, y) = c$, $x, y \in X$ is called totally constant.

Lemma 4 ([10]). *Let* $a, b \in [0, 1]$, $* : [0, 1]^2 \to [0, 1]$. *Totally constant fuzzy relations* T_a, T_b *have properties* $T_a \leq T_b \Leftrightarrow a \leq b, T_a = T_b \Leftrightarrow a = b, T_a \circ T_b = T_{a*b}$.

Theorem 1 ([10]). *Let* $* : [0, 1]^2 \to [0, 1]$.

- *Monotonicity of the operation* $*$ (*it is increasing or decreasing with respect to the first or to the second argument*) *is equivalent to suitable property of the composition* $\circ$.
- *The operation* $*$ *has* (*left, right*) *zero element* $z \in [0, 1]$ *if and only if the composition* $\circ$ *has suitable zero element* $Z = T_z$.
- *If the operation* $*$ *is increasing, then the composition* $\circ$ *is subdistributive over* $\wedge$, *i.e.*

$$T \circ (R \wedge S) \leqslant T \circ R \wedge T \circ S, \ (R \wedge S) \circ T \leqslant R \circ T \wedge S \circ T, \ R, S, T \in FR(X). \qquad (9)$$

- *The operation $*$ is (left, right) distributive over maximum if and only if the composition $\circ$ fulfils suitable distributivity property in $FR(X)$, i.e.*

$$T \circ (R \vee S) = T \circ R \vee T \circ S, \ (R \vee S) \circ T = R \circ T \vee S \circ T, \ R, S, T \in FR(X).$$

- *The operation $*$ is infinitely sup-distributive if and only if the composition $\circ$ is infinitely sup-distributive, i.e.*

$$R \circ (\sup_{t \in T} S_t) = \sup_{t \in T}(R \circ S_t), \quad (\sup_{t \in T} S_t) \circ R = \sup_{t \in T}(S_t \circ R),$$

where $R, S_t \in FR(X), t \in T$ for an arbitrary index set $T \neq \emptyset$.

- *Let the operation $*$ be infinitely sup-distributive. The operation $*$ is associative in $[0,1]$ if and only if the composition $\circ$ is associative in $FR(X)$.*
- *Let $z = 0$ be the zero element of the operation $*$. The operation $*$ has (left, right) neutral element $e \in (0,1]$ if and only if the composition $\circ$ has suitable neutral element $E \in FR(X)$, where*

$$E(x,y) = \begin{cases} e, & if \ x = y \\ 0, & otherwise \end{cases}, \quad x, y \in X. \qquad (10)$$

Directly from Theorem 1 and Lemma 3 we get

Corollary 3. *Let an operation $* : [0,1]^2 \to [0,1]$ be increasing.*

- *If the operation $*$ is left-continuous in $[0,1]$, then the composition $\circ$ is infinitely sup-distributive in $FR(X)$.*
- *If the operation $*$ is left-continuous and associative in $[0,1]$, then the composition $\circ$ is associative in $FR(X)$.*

We shall show, that the continuity assumption in this corollary cannot be omitted.

Example 3 ([10], Example 6). The drastic triangular norm $* = T_D$ (cf. (5)) is increasing and associative but it is not left-continuous. Let $R, S, U \in FR([0,1])$, where

$$R(x,y) = y, \quad U(x,y) = 1, \quad S(x,y) = \begin{cases} 0, & if \ \max(x,y) = 1 \\ y, & otherwise \end{cases}, \quad x, y \in [0,1].$$

By direct calculation we get $R \circ (S \circ U) = T_1 \neq (R \circ S) \circ U = T_0$ with totally constant relations T_1 and T_0. Thus the composition $\circ$ is not associative.

Remark 1. Under assumptions of the above corollary, the composition $\sup -*$ is associative and we can consider the sequence of powers of fuzzy relations: $R^1 = R$, $R^{n+1} = R^n \circ R$ for $n \in \mathbb{N}$. In the case of $* = \min$ the powers will be denoted by $R^{\bullet n}$.

4 Fuzzy Implications

Definitions and examples of fuzzy implications will be presented mainly after recent monograph by Baczyński and Jayaram [5].

Definition 5 ([5], p. 2). A function $I\colon [0,1]^2 \to [0,1]$ is called fuzzy implication if it is decreasing with respect to the first variable, increasing with respect to the second one and fulfils the binary implication truth table (cf. also (12)):

$$I(0,0) = I(1,1) = 1, \quad I(1,0) = 0. \tag{11}$$

The set of all fuzzy implications is denoted by FI.

Remark 2 ([5], p. 29). Let $I \in FI$. By reciprocal of I we call the following function $I'(x,y) = I(1-y, 1-x), x, y \in [0,1]$, which also is a fuzzy implication.

By monotonicity assumptions, from (11) we get

$$I(0,x) = I(x,1) = 1, \quad I(0,1) = 1, \quad x \in [0,1]. \tag{12}$$

First of all we can observe that the family FI is ordered by the relation (6) and we obtain

Theorem 2 ([2]). *The family FI is a convex set of functions. Moreover, $(FI, \vee, \wedge)$ is a complete, infinitely distributive lattice. In particular it has the least element I_0 and the greatest element I_1, where $(x,y \in [0,1])$*

$$I_0(x,y) = \begin{cases} 1, & \text{if } x = 0 \text{ or } y = 1 \\ 0, & \text{otherwise} \end{cases}, \ I_1(x,y) = \begin{cases} 0, & \text{if } x = 1 \text{ and } y = 0 \\ 1, & \text{otherwise} \end{cases}. \tag{13}$$

Example 4. The most important multivalued implications with theirs reciprocals (cf. [5], pp. 4, 30) fulfil the above definition $(x, y \in [0,1])$:

$$I_{LK}(x,y) = \min(1-x+y, 1), \quad I'_{LK} = I_{LK}, \quad \text{Łukasiewicz (1923)},$$

$$I_{GD}(x,y) = \begin{cases} 1, & \text{if } x \leqslant y \\ y, & \text{if } x > y \end{cases}, \quad I'_{GD}(x,y) = \begin{cases} 1, & \text{if } x \leqslant y \\ 1-x, & \text{if } x > y \end{cases}, \quad \text{Gödel (1932)},$$

$$I_{RC}(x,y) = 1-x+xy, \quad I'_{RC} = I_{RC}, \quad \text{Reichenbach (1935)},$$

$$I_{KD}(x,y) = \max(1-x, y), \quad I'_{KD} = I_{KD}, \quad \text{Kleene-Dienes (1938, 1949)},$$

$$I_{GG}(x,y) = \begin{cases} 1, & \text{if } x \leqslant y \\ \frac{y}{x}, & \text{if } x > y \end{cases}, \quad I'_{GG}(x,y) = \begin{cases} 1, & \text{if } x \leqslant y \\ \frac{1-x}{1-y}, & \text{if } x > y \end{cases}, \quad \text{Goguen (1969)},$$

$$I_{RS}(x,y) = \begin{cases} 1, & \text{if } x \leqslant y \\ 0, & \text{if } x > y \end{cases}, \quad I'_{RS} = I_{RS}, \quad \text{Rescher (1969)},$$

$$I_{YG}(x,y) = \begin{cases} 1, & \text{if } x = y = 0 \\ y^x, & \text{otherwise} \end{cases}, \quad I'_{YG}(x,y) = \begin{cases} 1, & \text{if } x = y = 1 \\ (1-x)^{1-y}, & \text{otherwise} \end{cases}, \quad \text{Yager (1980)},$$

$$I_{WB}(x,y) = \begin{cases} 1, & \text{if } x < 1 \\ y, & \text{if } x = 1 \end{cases}, \quad I'_{WB}(x,y) = \begin{cases} 1, & \text{if } y > 0 \\ 1-x, & \text{if } y = 0 \end{cases}, \quad \text{Weber (1983)},$$

$$I_{FD}(x,y) = \begin{cases} 1, & \text{if } x \leqslant y \\ \max(1-x,y), & \text{if } x > y \end{cases}, \quad I'_{FD} = I_{FD}, \quad \text{Fodor (1993)}.$$

5 Implication Compositions

Our approach is based on the observation that every $I \in FI$ can be considered as a fuzzy relation in $[0,1]$. Thus the composition $\circ$ (cf. (8) for $X = L = [0,1]$) can be applied to fuzzy implications. In general, such operation is not internal in FI. Using $* = \min$ we obtain $I_1 \circ I_1 = T_1$ (total constant), which does not fulfil (11). We need additional assumptions on the operation $*$ and used implications in order to obtain a result from FI. The case $* = \min$ we have examined in [3] and the case of triangular norm $*$ was examined in [4]. Similar results can be obtained under more general assumptions about the given operation $* : [0,1]^2 \to [0,1]$.

Lemma 5. *If the operation $*$ is increasing with respect to the first argument, then* $\sup -*$ *composition of fuzzy implications is a function decreasing with respect to the first variable.*

Proof. Let $I, J \in FI$, $K = I \circ J$,

$$K(x,z) = (I \circ J)(x,z) = \sup_{y \in [0,1]} (I(x,y) * J(y,z)), \quad x, z \in [0,1]. \tag{14}$$

If $x \leqslant u$, then $I(x,y) \geqslant I(u,y)$ by Definition 5 and

$$K(x,z) \geqslant \sup_{y \in [0,1]} (I(u,y) * J(y,z)) = K(u,z), \quad z \in [0,1],$$

i.e. the operation K is decreasing with respect to the first variable.

Similarly we get

Lemma 6. *If the operation $*$ is increasing with respect to the second argument, then* $\sup -*$ *composition of fuzzy implications is a function increasing with respect to the second variable.*

Lemma 7. *If the operation $*$ fulfils the condition $1 * 1 = 1$, then the composition (14) of fuzzy implications fulfils*

$$K(0,0) = K(1,1) = 1. \tag{15}$$

Proof. Let $I, J \in FI$, $K = I \circ J$. Directly from Definition 5 we get

$$1 \geqslant K(0,0) = \sup_{y \in [0,1]} (I(0,y) * J(y,0)) \geqslant I(0,0) * J(0,0) = 1 * 1 = 1,$$

$$1 \geqslant K(1,1) = \sup_{y \in [0,1]} (I(1,y) * J(y,1)) \geqslant I(1,1) * J(1,1) = 1 * 1 = 1,$$

which proves (15).

According to Lemmas 5, 6 and 7 we need only one condition from (11) in order to obtain $K \in FI$. Thus we get

Theorem 3 (cf. [4], Theorem 2). *If the operation $*$ is increasing and $1 * 1 = 1$, then for* $\sup -*$ *composition of fuzzy implications we have*

$$I \circ J \in FI \Leftrightarrow (I \circ J)(1,0) = 0. \tag{16}$$

Now, it must be provided, that the operation $*$ obtains the value 0.

Lemma 8. *Let $z = 0$ be the zero element of the operation $*$ and $I, J \in FI$.*

- *If $I(1,y) = 0$ for $y \in [0,1)$, then $(I \circ J)(1,0) = 0$.*
- *If $J(x,0) = 0$ for $x \in (0,1]$, then $(I \circ J)(1,0) = 0$.*

Proof. Let $I, J \in FI$. If $I(1,y) = 0$ for $y \in [0,1)$, then

$$(I \circ J)(1,0) = \sup_{y<1}(I(1,y) * J(y,0)) \vee (I(1,1) * J(1,0)) = 0 \vee 0 = 0.$$

If $J(y,0) = 0$ for $y \in (0,1]$, then

$$(I \circ J)(1,0) = (I(1,0) * J(0,0)) \vee \sup_{y>0}(I(1,y) * J(y,0)) = 0 \vee 0 = 0.$$

In both cases it proves the right hand condition from Theorem 3.

Lemma 8 leads to subfamilies of FI closed under the composition (14). We put

$$FI_L = \{I \in FI : \underset{x \in (0,1)}{\forall} I(1,x) = 0\}, \quad FI_R = \{I \in FI : \underset{x \in (0,1)}{\forall} I(x,0) = 0\}. \tag{17}$$

Additionally we put $FI_C = FI_L \cap FI_R$.

Remark 3. Let $I \in FI$. According to Remark 2 we get
$I \in FI_L \Leftrightarrow I' \in FI_R, I \in FI_R \Leftrightarrow I' \in FI_L, I \in FI_C \Leftrightarrow I' \in FI_C$.
Symbolically we have $FI'_L = FI_R, FI'_R = FI_L, FI'_C = FI_C$.

A comparison of (11) with (17) leads to

Corollary 4 ([3]). *All implications from $FI_L \cup FI_R$ are discontinuous.*

Example 5. Our conditions from (17) are restrictive and exclude many of presented examples of fuzzy implications (cf. (13) and Example 4). From Corollary 4 we know that every continuous fuzzy implication is excluded (e.g. I_{LK}, I_{RC} and I_{KD}). Moreover, by Remark 3 we have

$$I \notin FI_L \Leftrightarrow I' \notin FI_R, \ I \notin FI_R \Leftrightarrow I' \notin FI_L, \ I \notin FI_C \Leftrightarrow I' \notin FI_C.$$

Let $x \in (0,1)$. Since $I_1(x,0) = I_1(1,x) = I_{WB}(x,0) = 1$, $I_{FD}(x,0) = 1 - x$, $I_{GD}(1,x) = I_{GG}(1,x) = I_{YG}(1,x) = I_{WB}(1,x) = I_{FD}(1,x) = x$, then

$$I_1, I_{WB}, I'_{WB}, I_{FD} \notin FI_L \cup FI_R, \ I_{GD}, I_{GG}, I_{YG} \notin FI_L, \ I'_{GD}, I'_{GG}, I'_{YG} \notin FI_R.$$

Thus, as positive examples we get

$$I_0, I_{RS} \in FI_C, \ I_{GD}, I_{GG}, I_{YG} \in FI_R \setminus FI_L, \ I'_{GD}, I'_{GG}, I'_{YG} \in FI_L \setminus FI_R.$$

Corollary 5. *Families FI_L, FI_R and FI_C are complete, infinitely distributive sublattices of the lattice $(FI, \vee, \wedge)$ from Theorem 2. They are convex sets of functions bounded by suitable fuzzy implications:* $\min FI_L = \min FI_R = \min FI_C = I_0$, $\max FI_L = I_L$, $\max FI_R = I_R$, $\max FI_C = I_C$, *where*

$$I_L(x,y) = \begin{cases} 0, & \text{if } x = 1 \text{ and } y < 1 \\ 1, & \text{otherwise} \end{cases}, \ I_R(x,y) = \begin{cases} 0, & \text{if } x > 0 \text{ and } y = 0 \\ 1, & \text{otherwise} \end{cases}, \quad (18)$$

$$I_C(x,y) = \begin{cases} 0, & \text{if } (x,y) \in \{1\} \times [0,1) \cup (0,1] \times \{0\} \\ 1, & \text{otherwise} \end{cases}, \ x, y \in [0,1]. \quad (19)$$

Proof. Lattice operations (7) saves value 0 from (17). Thus families FI_L, FI_R and FI_C are closed with respect to arbitrary supremum and infimum. Then Theorem 2 implies infinite distributivity of sublattices $(FI_L, \vee, \wedge)$, $(FI_R, \vee, \wedge)$ and $(FI_C, \vee, \wedge)$. Since convex combination does not change constant values, then it saves conditions from (17). Thus the result belongs to the same family. Moreover, $I_0 \in FI_C$, thus the least implication is common. Formulas (18) are obtained from I_1 by minimal extension of $I^{-1}(\{0\})$ in order to obtain $I \in FI_L$ (FI_R). Finally, $I_C = I_L \wedge I_R$ which gives (19).

As a direct consequence of Lemma 8 and Theorem 3 we get

Corollary 6. *If the operation $*$ is increasing and $0*1 = 1*0 = 0$, $1*1 = 1$, then* $\sup -*$ *composition of fuzzy implications from families FI_L, FI_R and FI_C gives fuzzy implications.*

Using Lemma 2 we see that assumptions of Corollary 6 are fulfilled by arbitrary semicopula $*$ (cf. Definition 2). Returning to the condition (16) we have

Theorem 4 (cf. [4], Theorem 2). *If $*$ is a semicopula, then for* $\sup -*$ *composition of fuzzy implications we have*

$$I \in FI_L \Leftrightarrow \underset{J \in FI}{\forall} \ I \circ J \in FI, \quad J \in FI_R \Leftrightarrow \underset{I \in FI}{\forall} \ I \circ J \in FI. \tag{20}$$

Proof. Implications '⇒' are direct consequence of Lemma 8 and Theorem 3. Conversely, if $I \circ J \in FI$ for $J \in FI$, then putting $J = I_1$ we get

$$0 = (I \circ J)(1,0) \geqslant \sup_{x \in (0,1)} \ (I(1,x) * I_1(x,0)) = \sup_{x \in (0,1)} \ (I(1,x) * 1) = \sup_{x \in (0,1)} \ I(1,x).$$

Thus $I(1,x) = 0$ for $x \in (0,1)$, i.e. $I \in FI_L$, because $e = 1$ is the neutral element. Similarly we get $J \in FI_R$ in the second condition for $I = I_1$. This proves (20).

Lemma 9 (cf. [4], Lemma 2). *If* $0 * 1 = 1 * 0 = 0$, $1 * 1 = 1$ *and the operation* $*$ *is increasing, then families* FI_L, FI_R, FI_C *are closed with respect to the composition* (14) *and* $(FI_L, \circ, \leqslant)$, $(FI_R, \circ, \leqslant)$, $(FI_C, \circ, \leqslant)$ *are ordered groupoids.*

Proof. If $I, J \in FI_L$, $z \in (0,1)$, then using properties of the operation $*$ we get

$$(I \circ J)(1,z) = I(1,0) * J(0,z) \vee \sup_{y \in (0,1)} \ (I(1,y) * J(y,z)) \vee I(1,1) * J(1,z) = 0,$$

i.e. $I \circ J \in FI_L$. Similarly, for $I, J \in FI_R$, $x \in (0,1)$ we get

$$(I \circ J)(x,0) = I(x,0) * J(0,0) \vee \sup_{y \in (0,1)} \ (I(x,y) * J(y,0)) \vee I(x,1) * J(1,0) = 0,$$

i.e. $I \circ J \in FI_R$. Thus FI_C is also closed under $\circ$. According to Theorem 1 the composition $\circ$ is increasing and we obtain ordered groupoids.

If the operation $*$ is a semicopula, then from Theorem 1 we see that the relation composition has the neutral element E_1 (cf. (10)), but the identity relation is not an implication. The family of implications has its own neutral element of composition (8).

Theorem 5 (cf. [4]). *Let the operation* $*$ *be a semicopula.*

- *The Rescher implication* I_{RS} *is the neutral element of the composition* (14) *in* FI. *In particular,* I_{RS} *is the neutral element in groupoids* $(FI_L, \circ)$, $(FI_R, \circ)$ *and* $(FI_C, \circ)$.
- *The implication* I_0 *is the zero element in* $(FI_C, \circ)$, *a right zero in* $(FI_R, \circ)$, *and a left zero in* $(FI_L, \circ)$.

Proof. From Example 4 we see that the positive value $I_{RS}(x,y) = 1$ is only for $x \leqslant y$. Thus for arbitrary $I \in FI$, $x, z \in [0,1]$, by monotonicity properties of I we have

$$(I \circ I_{RS})(x,z) = \sup_{y \leq z} I(x,y) = I(x,z), \ (I_{RS} \circ I)(x,z) = \sup_{y \geq x} I(y,z) = I(x,z),$$

which proves that $I \circ I_{RS} = I_{RS} \circ I = I$. Similarly, the positive value $I_0(x,y) = 1$ is only for $x = 0$ or $y = 1$. Thus for arbitrary $I \in FI$ we get

$$(I_0 \circ I)(x,z) = I_0(x,1) * I(1,z) = I(1,z), \ (I \circ I_0)(x,z) = I(x,0) * I_0(0,z) = I(x,0).$$

Let $x \in (0,1]$, $z \in [0,1)$. If $I \in FI_L$, then $(I_0 \circ I)(x,z) = 0 = I_0(x,z)$. If $I \in FI_R$, then $(I \circ I_0)(x,z) = 0 = I_0(x,z)$. For $x = 0$ or $z = 1$ we can use the property (12) and we get $I_0 \circ I = I_0$ in FI_L, $I \circ I_0 = I_0$ in FI_R, which provides the zero element in FI_C.

The composition $\circ$ can be extended to subsets $A, B \subset FI$ by the formula $A \circ B = \{I \circ J | I \in A,\ J \in B\}$. Thus we can describe global properties of the composition $\circ$.

Theorem 6 (cf. [5], Theorem 6.4.11). *Let the operation $*$ be a semicopula. Considered families of fuzzy implications have the following properties*

$$FI_L \circ FI_L = FI_L,\ FI_R \circ FI_R = FI_R,\ FI_C \circ FI_C = FI_C, \tag{21}$$

$$FI_C \circ FI_R = FI_R \circ FI_C = FI_R,\ FI_C \circ FI_L = FI_L \circ FI_C = FI_L,\ FI_C \circ FI = FI \circ FI_C = FI, \tag{22}$$

$$I_1 \in FI_L \circ FI_R \subset FI_L \circ FI = FI \circ FI_R = FI,\ T_1 \in FI_R \circ FI_L \subset FI \circ FI_L \cap FI_R \circ FI \not\subset FI. \tag{23}$$

Proof. Properties (21) are consequences of Lemma 9 and Theorem 5, because $I_{RS} \in FI_C = FI_L \cap FI_R$. We also get

$$FI_R \subset FI_C \circ FI_R,\ FI_R \subset FI_R \circ FI_C,\ FI_L \subset FI_C \circ FI_L,\ FI_L \subset FI_L \circ FI_C, \tag{24}$$

$$FI \subset FI \circ FI_C \subset (FI \circ FI_L) \cap (FI \circ FI_R),\ FI \subset FI_C \circ FI \subset (FI_L \circ FI) \cap (FI_R \circ FI). \tag{25}$$

By direct calculation (cf. also Table 2, where $J_3 = I_R$, $J_4 = I_L$, $J_6 = I_C$) we get

$$I_L \circ I_L = I_L \circ I_C = I_C \circ I_L = I_L,\ I_R \circ I_R = I_R \circ I_C = I_C \circ I_R = I_R,\ I_C \circ I_C = I_C,$$

$$I_1 \circ I_R = I_1 \circ I_C = I_R \circ I_1 = I_C \circ I_1 = I_1,\ I_R \circ I_L = I_R \circ I_1 = I_1 \circ I_L = T_1.$$

According to Corollary 5 we have

$$FI_L = \{I \in FI | I \leqslant I_L\},\ FI_R = \{I \in FI | I \leqslant I_R\},\ FI_C = \{I \in FI | I \leqslant I_C\}.$$

Since the operation $*$ is increasing, then by Theorem 1 the composition $\circ$ is also increasing and we obtain

$$FI_C \circ FI_R = \{I \in FI | I \leqslant I_C \circ I_R = I_R\} \subset FI_R,$$

$$FI_R \circ FI_C = \{I \in FI | I \leqslant I_R \circ I_C = I_R\} \subset FI_R.$$

Similarly we get $FI_C \circ FI_L \subset FI_L$, $FI_L \circ FI_C \subset FI_L$, which, connected with (24) and (25), gives (22). Moreover, $FI_C \circ FI \subset FI$, $FI \circ FI_C \subset FI$ by Theorem 4. Finally,

$$I_1 = I_L \circ I_R \in FI_L \circ FI_R \subset FI_L \circ FI \subset FI,$$

$$T_1 = I_R \circ I_L \in FI_R \circ FI_L \subset (FI \circ FI_L) \cap (FI_R \circ FI)$$

and using (25) we obtain (23).

Remark 4. The above theorem disproves some inclusions suggested without proof in [4], Corollary 2, and [5], Theorem 6.4.11. In particular inclusions
$FI \circ FI_C \subset FI_C, FI_C \circ FI \subset FI_C, FI_C \circ FI_L \subset FI_C, FI \circ FI_L \subset FI_L, FI_R \circ FI \subset FI_R$
are false because of (22) and (23).

The most natural implication semigroups are semilattices $(FI, \vee, I_0)$ and $(FI, \wedge, I_1)$ from Theorem 2. These semigroups include all possible fuzzy implications from Definition 5. By Theorem 6 there are particular subfamilies in FI closed under sup$-*$ composition of fuzzy implications. Now we consider some additional assumptions in order to obtain ordered semigroups from ordered groupoids $(FI_L, \circ, \leqslant)$, $(FI_R, \circ, \leqslant)$ and $(FI_C, \circ, \leqslant)$. According to Theorem 1 and Corollary 3 we need an ordered monoid $([0,1], *, \leqslant)$ with left continuous operation $*$.

Theorem 7 (cf. [4], Theorem 3). *If* $([0,1], *, \leqslant)$ *is an ordered monoid with left continuous operation* $*$, *then algebraic structures* $(FI_L, \circ, I_{RS}, \leqslant)$, $(FI_R, \circ, I_{RS}, \leqslant)$ *and* $(FI_C, \circ, I_{RS}, \leqslant)$ *are also ordered monoids, where the composition* sup$-*$ *is infinitely sup-distributive. Moreover, the monoid* $(FI_C, \circ, I_{RS}, \leqslant)$ *has the zero element* I_0.

Proof. In virtue of Lemma 9 and Theorem 5 we know that algebraic structures $(FI_L, \circ, I_{RS}, \leqslant)$, $(FI_R, \circ, I_{RS}, \leqslant)$ and $(FI_C, \circ, I_{RS}, \leqslant)$ are ordered groupoids with the neutral element I_{RS}. Now, by Corollary 3, the sup$-*$ composition is associative and infinitely sup-distributive. Thus algebraic structures $(FI_L, \circ, I_{RS}, \leqslant)$, $(FI_R, \circ, I_{RS}, \leqslant)$ and $(FI_C, \circ, I_{RS}, \leqslant)$ are ordered semigroups with neutral element, i.e. ordered monoids. As usual, the least element I_0 in such monoids is the zero element in the lower subsemigroup $\{I \in FI | I \leqslant I_{RS}\} \subset FI_C$, but by Theorem 5 it is the zero element in $(FI_C, \circ, I_{RS})$.

Corollary 7 (cf. [5], Theorem 6.4.13). *If the operation* $*$ *is left continuous triangular norm, pseudo triangular norm or conjunctive uninorm then it fulfils assumptions of Theorem 7. Therefore, algebraic structures* $(FI_L, \circ, I_{RS}, \leqslant)$, $(FI_R, \circ, I_{RS}, \leqslant)$, $(FI_C, \circ, I_{RS}, \leqslant)$ *are ordered monoids with infinitely sup-distributive composition* sup$-*$ *and the monoid* $(FI_C, \circ, I_{RS}, \leqslant)$ *has the zero element* I_0.

Under assumptions of the above theorem one can consider the sequence of powers of fuzzy implications (cf. Remark 1).

Example 6. Since I_{GG} belongs to FI_R and $* = \min$ fulfils assumptions of Theorem 7, then we can consider powers of I_{GG}. At first we get:
• if $x \leqslant z$, then

$$(I_{GG} \bullet I_{GG})(x,z) = \sup_{y \in [0,1]} \min(I_{GG}(x,y), I_{GG}(y,z)) \geqslant \min(I_{GG}(x,x), I_{GG}(x,z)) = 1;$$

• if $x > z$, then

$$(I_{GG} \bullet I_{GG})(x,z) = \sup_{\substack{y \in [0,1]}} \min(I_{GG}(x,y),I_{GG}(y,z)) = \sup_{\substack{y \in [0,1] \\ y \leqslant z < x}} \min(I_{GG}(x,y),I_{GG}(y,z))$$

$$\vee \sup_{\substack{y \in [0,1] \\ z < y < x}} \min(I_{GG}(x,y),I_{GG}(y,z)) \vee \sup_{\substack{y \in [0,1] \\ z < x \leqslant y}} \min(I_{GG}(x,y),I_{GG}(y,z))$$

$$= \sup_{\substack{y \in [0,1] \\ y \leqslant z < x}} \min\left(\frac{y}{x},1\right) \vee \sup_{\substack{y \in [0,1] \\ z < y < x}} \min\left(\frac{y}{x},\frac{z}{y}\right) \vee \sup_{\substack{y \in [0,1] \\ z < x \leqslant y}} \min\left(1,\frac{z}{y}\right) = \frac{z}{x} \vee \sqrt{\frac{z}{x}} = \sqrt{\frac{z}{x}}.$$

Thus by mathematical induction we obtain

$$I_{GG}^{\bullet n}(x,y) = \begin{cases} 1, & \text{if } x \leqslant y \\ \sqrt[n]{\frac{y}{x}}, & \text{if } x > y \end{cases}, \tag{26}$$

where all these powers belong to FI_R. Similarly, I'_{GG} belongs to FI_L and we obtain

$$(I'_{GG})^{\bullet n}(x,y) = \begin{cases} 1, & \text{if } x \leqslant y \\ \sqrt[n]{\frac{1-x}{1-y}}, & \text{if } x > y \end{cases}, \tag{27}$$

where all these powers belong to FI_L. Finally, $I_{GG} \wedge I'_{GG}$ belongs to FI_L, where

$$(I_{GG} \wedge I'_{GG})(x,y) = \begin{cases} 1, & \text{if } x \leqslant y \\ \frac{y}{x}, & \text{if } y < x \leqslant 0.5 \\ \frac{1-x}{1-y}, & \text{if } 0.5 < y < x \end{cases}.$$

According to (9) we have $(I_{GG} \wedge I'_{GG})^n \leqslant K_n = I_{GG}^{\bullet n} \wedge (I'_{GG})^{\bullet n}$, where

$$K_n(x,y) = \begin{cases} 1, & \text{if } x \leqslant y \\ \sqrt[n]{\frac{y}{x}}, & \text{if } y < x \leqslant 0.5 \\ \sqrt[n]{\frac{1-x}{1-y}}, & \text{if } 0.5 < y < x \end{cases} \tag{28}$$

and implications K_n belong to FI_C. Let us observe that

$$\sup I_{GG}^{\bullet n} = I_R, \ \sup(I'_{GG})^{\bullet n} = I_L, \ \sup K_n = I_C.$$

Because of Corollary 2 we get

Corollary 8. *If a left continuous operation $*$ is a pseudo triangular norm, then*

$$I_{GG}^n \leqslant I_{GG}^{\bullet n}, \ (I'_{GG})^n \leqslant (I'_{GG})^{\bullet n}, \ (I_{GG} \wedge I_{GG})^n \leqslant I_{GG}^{\bullet n} \wedge (I'_{GG})^{\bullet n}.$$

6   Compositions in Classes of Fuzzy Implications

Now we consider some important additional properties of fuzzy implications.

Definition 6 ([5], pp. 9, 20). A fuzzy implication I is said to satisfy:

(NP), the left neutral property, if	$I(1,y) = y,\ y \in [0,1]$,
(EP), the exchange principle, if	$I(x,I(y,z)) = I(y,I(x,z)),\ x,y,z \in [0,1]$,
(IP), the identity principle, if	$I(x,x) = 1,\ x \in [0,1]$,
(OP), the ordering property, if	$I(x,y) = 1 \Leftrightarrow x \leqslant y,\ x,y \in [0,1]$,
(CP), the law of contraposition, if	$I(x,y) = I(1-y,1-x),\ x,y \in [0,1]$.

Families of fuzzy implications determined by the above properties will be denoted by $NP, EP, IP, OP, CP \subset FI$, respectively.

Table 1 Main properties of basic fuzzy implications

Case	(NP)	(EP)	(IP)	(OP)	(CP)
I_0	—	+	—	—	+
I_1	—	+	+	—	+
I_{LK}	+	+	+	+	+
I_{GD}	+	+	+	+	—
I'_{GD}	—	—	+	+	—
I_{RC}	+	+	—	—	+
I_{KD}	+	+	—	—	+
I_{GG}	+	+	+	+	—
I'_{GG}	—	—	+	+	—
I_{RS}	—	—	+	+	+
I_{YG}	+	+	—	—	—
I'_{YG}	—	—	—	—	—
I_{WB}	+	+	+	—	—
I'_{WB}	—	—	+	—	—
I_{FD}	+	+	+	+	+

Example 7. It can be verified (cf. [5], pp. 10, 29) that the fuzzy implications listed in Example 4 fulfil some of additional properties described in the above definition, what is presented on Table 1, where symbol '+' means that the function from the chosen row has the property from the chosen column, and symbol '−' means that there exists a counter-example. Since implications I_{GD}, I_{GG}, I_{YG} and I_{WB} are not contrapositive, then also their reciprocals I'_{GD}, I'_{GG}, I'_{YG}, $I'_{WB} \in FI$ do not fulfil (CP). Properties (IP), (OP) are preserved from main versions by reciprocals (cf. [5], p. 30) and the property (NP) can be simply verified. We shall show that the property (EP) is not fulfilled. Let $x = 0.5$, $y = 1$, $z = 0$, $L(I) = I(x,I(y,z))$, $R(I) = I(y,I(x,z))$. Directly from the above formulas we get $R(I'_{GD}) = R(I'_{GG}) = R(I'_{YG}) = 0$, $R(I'_{WB}) = 1$, $L(I'_{GD}) = L(I'_{GG}) = L(I'_{YG}) = L(I'_{WB}) = 0.5$, This is contradictory with (EP).

Now we can consider new sublattices of the lattice $(FI, \vee, \wedge)$.

Theorem 8 ([5], pp.185-187). *Ordered structures* $(IP, \vee, \wedge)$, $(OP, \vee, \wedge)$, $(NP, \vee, \wedge)$, $(CP, \vee, \wedge)$ *are distributive lattices and convex families of functions. While family* EP *is not closed with respect to lattice operations and convex combinations.*

If we investigate composition of fuzzy implication in classes NP, EP, IP, OP and CP we must apply Theorems 4 and 6 and use implications from families (17), otherwise the result need not to be an implication (e.g. $I_1 \in IP$ but $I_1 \circ I_1 = T_1 \notin IP$).

Lemma 10. *The Rescher implication* I_{RS} *is the least element of IP and OP. Moreover* $\max IP = \sup OP = I_1$.

Proof. Simply we get $I_{RS} \in IP \cap OP$. Now, if $I < I_{RS}$, then there exist $x, y \in [0, 1]$, such that $x \leqslant y$ and $I(x, y) < 1$. Thus $I \notin OP$. Since $I(x, x) \leqslant I(x, y) < 1$ by Definition 5, then $I \notin IP$. Thus $\min IP = \min OP = I_{RS}$.

Moreover, $\sup OP \leqslant \sup IP \leqslant \max FI = I_1$. Since $I_1 \in IP$, then $\max IP = I_1$. However $I_1 \notin OP$, but there exists a sequence $L_n \in OP, n \in \mathbb{N}$, such that $\sup L_n = I_1$, e.g.

$$L_n(x, y) = \begin{cases} 1, & \text{if } (x \leqslant y) \\ 0, & \text{if } (x, y) = (1, 0), \ x, y \in [0, 1]. \\ 1 - \frac{1}{n}, & \text{otherwise} \end{cases}$$

Thus $\sup OP = I_1$.

Next lemma concerns fuzzy relations $I, J \in FR([0, 1])$.

Lemma 11. *Let the operation* $*$ *be a semicopula and* $I, J \in FR([0, 1])$.
• *If* I, J *fulfil (IP), then* $I \circ J$ *fulfils (IP).*
• *If* I, J *fulfil (OP), then* $I \circ J$ *fulfils (OP).*

Proof. Let $x \in [0, 1]$ and I, J fulfil (IP). Thus $I(x, x) = J(x, x) = 1$, and we obtain

$$1 \geqslant (I \circ J)(x, x) = \sup_{y \in [0,1]} I(x, y) * J(y, x) \geqslant I(x, x) * J(x, x) = 1,$$

which proves that $(I \circ J)(x, x) = 1$, i.e. $I \circ J$ fulfils (IP). Now let $x, z \in [0, 1]$, $x \leqslant z$. If I, J fulfil (OP), then

$$1 \geqslant (I \circ J)(x, z) = \sup_{y \in [0,1]} I(x, y) * J(y, z) \geqslant I(x, x) * J(x, z) = 1,$$

which proves that $(I \circ J)(x, z) = 1$. Conversely, suppose that there exist $x, z \in [0, 1]$, $x > z$, such that

$$1 = (I \circ J)(x, z) = \sup_{\substack{y \in [0,1] \\ y \leqslant z < x}} I(x, y) * J(y, z) \vee \sup_{\substack{y \in [0,1] \\ z < y < x}} I(x, y) * J(y, z) \vee \sup_{\substack{y \in [0,1] \\ z < x \leqslant y}} I(x, y) * J(y, z)$$

$$= \sup_{\substack{y \in [0,1] \\ y \leqslant z < x}} I(x, y) * 1 \vee \sup_{\substack{y \in [0,1] \\ z < y < x}} I(x, y) * J(y, z) \vee \sup_{\substack{y \in [0,1] \\ z < x \leqslant y}} 1 * J(y, z)$$

$$= I(x,z) \vee \sup_{\substack{y\in[0,1]\\z<y<x}} I(x,y) * J(y,z) \vee J(x,z).$$

Since $I(x,z) < 1$ and $J(x,z) < 1$, then we get

$$\sup_{\substack{y\in[0,1]\\z<y<x}} I(x,y) * J(y,z) = 1$$

and there exists an infinite sequence $y_n \in (z,x)$, $n \in \mathbb{N}$, such that

$$I(x,y_n) > 1 - \frac{1}{n}, \quad J(y_n,z) > 1 - \frac{1}{n}, \quad n \in \mathbb{N}. \tag{29}$$

Such bounded sequence contains a convergent subsequence $t_k = y_{n_k}$, $k \in \mathbb{N}$, and one can choose it strictly monotonic. If $t_{k+1} < t_k$ for $k \in \mathbb{N}$, then $I(x,t_{k+1}) \leqslant I(x,t_k) \leqslant \dots \leqslant I(x,y_1) < 1$. Similarly, if $t_{k+1} > t_k$ for $k \in \mathbb{N}$, then $J(t_{k+1},z) \leqslant J(t_k,z) \leqslant \dots \leqslant J(y_1,z) < 1$, contrary to (29). Thus $(I \circ J)(x,z) < 1$, for $x > z$, i.e. $I \circ J$ fulfils (OP).

Directly from Lemmas 9, 10, 11 and Theorem 7 we obtain

Theorem 9. *Let $IP_L = IP \cap FI_L$, $IP_R = IP \cap FI_R$, $IP_C = IP \cap FI_C$.*

- *If $*$ is a semicopula, then $(IP_L,\circ,I_{RS},\leqslant)$, $(IP_R,\circ,I_{RS},\leqslant)$, $(IP_C,\circ,I_{RS},\leqslant)$ are ordered groupoids with the neutral element I_{RS}. Moreover, the above families are lattice intervals in FI: $IP_L = [I_{RS},I_L]$, $IP_R = [I_{RS},I_R]$, $IP_C = [I_{RS},I_C]$.*
- *If $*$ is a left continuous pseudo t-norm, then $(IP_L,\circ,I_{RS},\leqslant)$, $(IP_R,\circ,I_{RS},\leqslant)$ and $(IP_C,\circ,I_{RS},\leqslant)$ are ordered monoids.*

Theorem 10. *Let $OP_L = OP \cap FI_L$, $OP_R = OP \cap FI_R$, $OP_C = OP \cap FI_C$.*

- *If $*$ is a semicopula, then $(OP_L,\circ,I_{RS},\leqslant)$, $(OP_R,\circ,I_{RS},\leqslant)$, $(OP_C,\circ,I_{RS},\leqslant)$ are ordered groupoids with the neutral element I_{RS}.*
- *If $*$ is a left continuous pseudo t-norm, then $(OP_L,\circ,I_{RS},\leqslant)$, $(OP_R,\circ,I_{RS},\leqslant)$ and $(OP_C,\circ,I_{RS},\leqslant)$ are ordered monoids.*

In the case of (OP) the above families are not closed lattice intervals but they contain some intervals.

Example 8. By Example 6 (case $* = \min$) one can check that $I_{GG}^{\bullet n} \in OP \cap FI_R$, $(I'_{GG})^{\bullet n} \in OP \cap FI_L$ and $K_n \in OP \cap FI_C$. Thus $[I_{RS},I_{GG}^{\bullet n}] \subset OP \cap FI_R$, $[I_{RS},(I'_{GG})^{\bullet n}] \subset OP \cap FI_L$, $[I_{RS},I_{GG}^{\bullet n} \wedge (I'_{GG})^{\bullet n}] \subset OP \cap FI_C$ for $n \in \mathbb{N}$ (cf. Lemma 10). In general, if a left continuous operation $*$ is a pseudo triangular norm, then by Corollary 8 we also have
$I_{GG}^n \in OP \cap FI_R$, $(I'_{GG})^n \in OP \cap FI_L$, $(I_{GG} \wedge I'_{GG})^n \in OP \cap FI_C$ for $n \in \mathbb{N}$.

From the condition (NP) we obtain $NP \cap FI_L = NP \cap FI_C = \emptyset$. So we put attention only on the family FI_R. However, $I_{GG} \in NP \cap FI_R$, and from Example 6 we get $I_{GG}^{\bullet 2}(1,y) = \sqrt{y}$, $y \in [0,1]$. Thus $I_{GG}^{\bullet 2} \notin NP$ and the composition $\bullet$ is not internal operation in $NP \cap FI_R$.

Families (17) have common part with CP. Directly from Remark 3 we get

Remark 5. We have $CP \cap FI_L = CP \cap FI_R = CP \cap FI_C$.

Theorem 11. *Let the operation $*$ be a commutative semicopula. If fuzzy implications $I, J \in CP \cap FI_C$ commutes, i. e. $I \circ J = J \circ I$, then $I \circ J \in CP \cap FI_C$.*

Proof. Let $I, J \in CP \cap FI_C$, $x, z \in [0,1]$. By Lemma 9 we have $I \circ J \in FI_C$. Using (CP) and commutativity of $*$ we obtain

$$
\begin{aligned}
(I \circ J)(1 - z, 1 - x) &= \sup_{y \in [0,1]} I(1 - z, y) * J(y, 1 - x) \\
&= \sup_{y \in [0,1]} I(1 - z, 1 - y) * J(1 - y, 1 - x) \\
&= \sup_{y \in [0,1]} I(y, z) * J(x, y) = \sup_{y \in [0,1]} J(x, y) * I(y, z) = (J \circ I)(x, z).
\end{aligned}
$$

Since implications I, J commutes, then we obtain $I \circ J \in CP$, what finishes the proof.

Case $I = J$ leads to the following (cf. Corollary 7)

Corollary 9. *If the operation $*$ is a left continuous triangular norm, then family $CP \cap FI_C$ is closed with respect to powers of its own elements.*

Example 9. Composition of elements from EP need not be from EP. Let $x = 0.64$, $y = 1$, $z = 0.49$ and $I = I_{GG}^{\bullet 2}$, where $I_{GG} \in EP \cap FI_R$. Since $I(0.64, I(1, 0.49)) = I(0.64, 0.7) = 1$, $I(1, I(0.64, 0.49)) = I(1, \frac{7}{8}) = \sqrt{\frac{7}{8}} < 1$, then $I \notin EP \cap FI_R$.

7 Invariant Fuzzy Implications

Now we shall consider a connection between families FI_L, FI_R, FI_C and conjugacy classes of fuzzy implications.

Definition 7 ([1]). Let Φ denote the set of all increasing bijections on $[0,1]$. Fuzzy implications $I, J \in FI$ are conjugate if there exists $\varphi \in \Phi$ such that $I_\varphi = J$, where

$$I_\varphi(x, y) = \varphi^{-1}(I(\varphi(x), \varphi(y))), \qquad x, y \in [0,1]. \tag{30}$$

A fuzzy implication $I \in FI$ is invariant with respect to Φ (selfconjugate) if

$$\underset{\varphi \in \Phi}{\forall} \, I_\varphi = I, \tag{31}$$

The family of all invariant fuzzy implications is denoted by IFI.

Example 10. The following fuzzy implications I_{GD}, I_{RS}, I_0, I_1, I_L, I_R and I_C belong to IFI (cf. [11]).

Theorem 12 ([3]). *Families FI_L, FI_R and FI_C are closed under operation (30).*

Proof. If $I \in FI_L$, $\varphi \in \Phi$, $z \in [0,1]$ then

$$I_\varphi(1,z) = \varphi^{-1}(I(\varphi(1),\varphi(z))) = \varphi^{-1}(I(1,\varphi(z))) = \varphi^{-1}(0) = 0,$$

i.e. $I_\varphi \in FI_L$. Similar calculation is valid for $I \in FI_R$, and $I \in FI_C$ fulfils both properties ($I_\varphi \in FI_L \cap FI_R$).

Lemma 12. *The transformation* (30) *is distributive with respect to* $\sup - \min$ *composition, i.e.*

$$(I \bullet J)_\varphi = I_\varphi \bullet J_\varphi, \ J \in FI, \ \varphi \in \Phi. \tag{32}$$

Proof. Let $I, J \in FI$, $\varphi \in \Phi$, $x, z \in [0,1]$. We have

$$(I \bullet J)_\varphi(x,z) = \varphi^{-1}((I \bullet J)(\varphi(x),\varphi(z))) = \varphi^{-1}(\sup_{y \in [0,1]} I(\varphi(x),y) \wedge J(y,\varphi(z)))$$

$$= \varphi^{-1}(\sup_{y \in [0,1]} I(\varphi(x),\varphi(y)) \wedge J(\varphi(y),\varphi(z)))$$

$$= \sup_{y \in [0,1]} \varphi^{-1}(I(\varphi(x),\varphi(y))) \wedge \varphi^{-1}(J(\varphi(y),\varphi(z)))$$

$$= \sup_{y \in [0,1]} I_\varphi(x,y) \wedge J_\varphi(y,z) = (I_\varphi \bullet J_\varphi)(x,z),$$

which proves (32).

Theorem 13. *The* $\sup - \min$ *composition of invariant fuzzy implications is an invariant function. It is an invariant implication under condition from Theorem 3.*

Proof. If fuzzy implications $I, J \in FI$ are invariant, then directly from (31) and (32) we get $(I \bullet J)_\varphi = I_\varphi \bullet J_\varphi = I \bullet J$ for $\varphi \in \Phi$, i.e. $I \circ J$ is invariant.

The family IFI of all invariant fuzzy implications was examined in [11], [12] and we summarize some results.

Theorem 14 ([12], Theorem 6). *The family IFI of all invariant fuzzy implications is finite and consists of 18 implications $J_1, \ldots, J_{18}$, where:* $J_1 = I_1$, $J_2 = I_{WB}$, $J_3 = I_R$, $J_4 = I_L$, $J_6 = I_C$, $J_7 = I_{GD}$, $J_{10} = I_{RS}$, $J_{18} = I_0$,

$$J_8(x,y) = \begin{cases} 1, & \text{if } x \leq y \\ y, & \text{if } y < x < 1, \\ 0, & \text{otherwise} \end{cases} \qquad J_5(x,y) = \begin{cases} 1, & \text{if } (x < 1) \wedge (y > 0) \vee (x = 0) \\ y, & \text{if } x = 1 \\ 0, & \text{otherwise} \end{cases},$$

$$J_9(x,y) = \begin{cases} 1, & \text{if } (x < y) \vee (x = 0) \\ y, & \text{otherwise} \end{cases}, \qquad J_{11}(x,y) = \begin{cases} 1, & \text{if } (x < y) \vee (x = 0) \vee (y = 1) \\ y, & \text{if } (y \leq x) \wedge (0 < x < 1) \\ 0, & \text{otherwise} \end{cases},$$

$$J_{12}(x,y) = \begin{cases} 1, & \text{if } x = 0 \\ y, & \text{otherwise} \end{cases}, \qquad J_{13}(x,y) = \begin{cases} 1, & \text{if } (x < y) \vee (x = 0) \\ y, & \text{if } x = y > 0 \\ 0, & \text{otherwise} \end{cases},$$

$$J_{14}(x,y) = \begin{cases} 1, & \text{if } (x=0) \vee (y=1) \\ y, & \text{if } 0 < x < 1 \\ 0, & \text{otherwise} \end{cases}, \quad J_{15}(x,y) = \begin{cases} 1, & \text{if } (x<y) \vee (x=0) \vee (y=1) \\ 0, & \text{otherwise} \end{cases},$$

$$J_{16}(x,y) = \begin{cases} 1, & \text{if } x=0 \\ y, & \text{if } 0 < x \le y \\ 0, & \text{otherwise} \end{cases}, \quad J_{17}(x,y) = \begin{cases} 1, & \text{if } (x=0) \vee (y=1) \\ y, & \text{if } 0 < x < y < 1 \\ 0, & \text{otherwise} \end{cases}$$

for $x,y \in [0,1]$.

Theorem 15 ([11], Theorems 7, 8). *Invariant fuzzy implications $J_1,\ldots,J_{18}$ form the distributive lattice presented in Fig. 1 and their reciprocals form analogical Hasse diagram.*

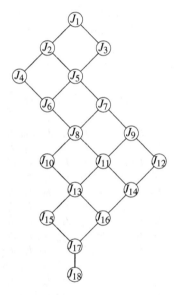

Fig. 1 The lattice of invariant fuzzy implications $J_1,\ldots,J_{18}$

Remark 6. By Remark 2 we obtain new fuzzy implications: $J_2' = I_{WB}'$, $J_7' = I_{GD}'$,

$$J_5'(x,y) = \begin{cases} 1, & \text{if } (x<1) \wedge (y>0) \vee (y=1) \\ 1-x, & \text{if } y=0 \\ 0, & \text{otherwise} \end{cases}, \quad J_8'(x,y) = \begin{cases} 1, & \text{if } x \le y \\ 1-x, & \text{if } 0 < y < x, \\ 0, & \text{otherwise} \end{cases}$$

$$J_9'(x,y) = \begin{cases} 1, & \text{if } (x<y) \vee (y=1) \\ 1-x, & \text{otherwise} \end{cases}, \quad J_{12}'(x,y) = \begin{cases} 1, & \text{if } y=1 \\ 1-x, & \text{otherwise} \end{cases},$$

$$J'_{11}(x,y) = \begin{cases} 1, & \text{if } (x<y)\vee(x=0)\vee(y=1) \\ 1-x, & \text{if } (y\le x)\wedge(0<y<1) \\ 0, & \text{otherwise} \end{cases} , \quad J'_{13}(x,y) = \begin{cases} 1, & \text{if } (x<y)\vee(y=1) \\ 1-x, & \text{if } x=y<1 \\ 0, & \text{otherwise} \end{cases} ,$$

$$J'_{14}(x,y) = \begin{cases} 1, & \text{if } (x=0)\vee(y=1) \\ 1-x, & \text{if } 0<y<1 \\ 0, & \text{otherwise} \end{cases} , \quad J'_{16}(x,y) = \begin{cases} 1, & \text{if } y=1 \\ 1-x, & \text{if } x\le y<1, \\ 0, & \text{otherwise} \end{cases}$$

$$J'_{17}(x,y) = \begin{cases} 1, & \text{if } (x=0)\vee(y=1) \\ 1-x, & \text{if } 0<x<y<1 \\ 0, & \text{otherwise} \end{cases}$$

for $x,y \in [0,1]$.

Compositions of invariant fuzzy implications were examined for $* = \min$.

Theorem 16 ([12], Theorem 6). *Let $J_0(x,y)=1$, $x,y \in [0,1]$. The composition table of invariant fuzzy implications with* $\sup - \min$ *composition has the form presented in Table 2.*

Table 2 Composition table of invariant fuzzy implications

•	J_1	J_2	J_3	J_4	J_5	J_6	J_7	J_8	J_9	J_{10}	J_{11}	J_{12}	J_{13}	J_{14}	J_{15}	J_{16}	J_{17}	J_{18}
J_1	J_0	J_0	J_1	J_0	J_1	J_1	J_1	J_1	J_1	J_1	J_1	J_2	J_1	J_2	J_1	J_2	J_2	J_4
J_2	J_0	J_0	J_1	J_0	J_1	J_1	J_2	J_2	J_2	J_2	J_2	J_2	J_2	J_2	J_2	J_2	J_2	J_4
J_3	J_0	J_0	J_3	J_0	J_3	J_3	J_3	J_3	J_3	J_3	J_3	J_{12}	J_3	J_{12}	J_3	J_{12}	J_{12}	J_{18}
J_4	J_1	J_2	J_1	J_4	J_2	J_4	J_2	J_4	J_2	J_4	J_4	J_2	J_4	J_4	J_4	J_4	J_4	J_4
J_5	J_0	J_0	J_3	J_0	J_3	J_3	J_5	J_5	J_5	J_5	J_5	J_{12}	J_5	J_{12}	J_5	J_{12}	J_{12}	J_{18}
J_6	J_1	J_2	J_3	J_4	J_5	J_6	J_5	J_6	J_5	J_6	J_6	J_{12}	J_6	J_{14}	J_6	J_{14}	J_{14}	J_{18}
J_7	J_0	J_0	J_3	J_0	J_3	J_3	J_7	J_7	J_9	J_7	J_9	J_{12}	J_9	J_{12}	J_9	J_{12}	J_{12}	J_{18}
J_8	J_1	J_2	J_3	J_4	J_5	J_6	J_7	J_8	J_9	J_8	J_{11}	J_{12}	J_{11}	J_{14}	J_{11}	J_{14}	J_{14}	J_{18}
J_9	J_0	J_0	J_3	J_0	J_3	J_3	J_9	J_9	J_9	J_9	J_9	J_{12}	J_9	J_{12}	J_9	J_{12}	J_{12}	J_{18}
J_{10}	J_1	J_2	J_3	J_4	J_5	J_6	J_7	J_8	J_9	J_{10}	J_{11}	J_{12}	J_{13}	J_{14}	J_{15}	J_{16}	J_{17}	J_{18}
J_{11}	J_1	J_2	J_3	J_4	J_5	J_6	J_9	J_{11}	J_9	J_{11}	J_{11}	J_{12}	J_{11}	J_{14}	J_{11}	J_{14}	J_{14}	J_{18}
J_{12}	J_0	J_0	J_3	J_0	J_3	J_3	J_{12}	J_{12}	J_{12}	J_{12}	J_{12}	J_{12}	J_{12}	J_{12}	J_{12}	J_{12}	J_{12}	J_{18}
J_{13}	J_1	J_2	J_3	J_4	J_5	J_6	J_9	J_{11}	J_9	J_{13}	J_{11}	J_{12}	J_{13}	J_{14}	J_{15}	J_{16}	J_{17}	J_{18}
J_{14}	J_1	J_2	J_3	J_4	J_5	J_6	J_{12}	J_{14}	J_{12}	J_{14}	J_{14}	J_{12}	J_{14}	J_{14}	J_{14}	J_{14}	J_{14}	J_{18}
J_{15}	J_1	J_2	J_3	J_4	J_5	J_6	J_9	J_{11}	J_9	J_{15}	J_{11}	J_{12}	J_{15}	J_{14}	J_{15}	J_{17}	J_{17}	J_{18}
J_{16}	J_1	J_2	J_3	J_4	J_5	J_6	J_{12}	J_{14}	J_{12}	J_{16}	J_{14}	J_{12}	J_{16}	J_{14}	J_{17}	J_{16}	J_{17}	J_{18}
J_{17}	J_1	J_2	J_3	J_4	J_5	J_6	J_{12}	J_{14}	J_{12}	J_{17}	J_{14}	J_{12}	J_{17}	J_{14}	J_{17}	J_{17}	J_{17}	J_{18}
J_{18}	J_3	J_{12}	J_3	J_{18}	J_{12}	J_{18}	J_{12}	J_{18}	J_{12}	J_{18}	J_{18}	J_{12}	J_{18}	J_{18}	J_{18}	J_{18}	J_{18}	J_{18}

Directly from Table 2 we see that

Corollary 10. *The set* $\{J_0,\ldots,J_{18}\}$ *is a semigroup with zero* J_0 *and neutral element* J_{10}.

The above theorem provides additional examples of semigroups of fuzzy implications. Directly from Theorems 7 and 16 we get

Corollary 11. *The following subsets of IFI are examples of finite semigroups of fuzzy implications:*

$$IFI \cap FI_R = \{J_3, J_5, \ldots, J_{18}\}, \ IFI \cap FI_C = \{J_6, J_8, J_{10}, J_{11}, J_{13}, \ldots, J_{18}\},$$

$$IFI \cap FI_L = IFI \cap FI_C \cup \{J_4\}.$$

Moreover, fuzzy implications $J_7, \ldots, J_{17}$ *form a commutative semigroup with the zero* J_{12} *and the neutral element* J_{10}.

8 Conclusion

Fuzzy implications play a central role in many inference schemes in approximate reasoning. The examination of compositions of fuzzy implications is very important not only in fuzzy logic but also in multistage decision making and fuzzy control.

The above results precise conditions for associativity of implication compositions which is necessary in multiple repetition of implications (e.g. in multistage decision making). We see, that used triangular norm $*$ cannot be arbitrary but assumption on left-continuity suffice for "regular" properties of composition (8). In particular, the most commonly used triangular norms (cf. Example 2) are continuous which guarantee semigroup properties of composition (8) (cf. Corollary 7). Finally, compositions of invariant fuzzy implications give a nice example of algebraic properties in fuzzy logic. For example, idempotent fuzzy implications from Table 2 have a stability property: a repetition does not change the result.

The above examination can be continued in diverse directions. First of all the dual results concerning $\inf{-}*$ composition are examined in connection with interval-valued fuzzy sets and Atanassov's intuitionistic fuzzy sets. Next, the interval $[0,1]$ of truth values can be replaced by certain lattice L (L-fuzzy implications). Finally, there exist other relation compositions which can be used for fuzzy implications (cf. e.g. BK-compositions introduced by Bandler and Kohout [7] with results summarized by Bělohlávek [8], Chapter 6).

Acknowledgements. This work is partially supported by the Ministry of Science and Higher Education Grant Nr N N519 384936 (Poland).

References

1. Baczyński, M., Drewniak, J.: Conjugacy Classes of Fuzzy Implication. In: Reusch, B. (ed.) Fuzzy Days 1999. LNCS, vol. 1625, pp. 287–298. Springer, Heidelberg (1999)
2. Baczyński, M., Drewniak, J.: Monotonic Fuzzy Implication. In: Szczepaniak, P.S., Lisboa, P.J.G., Kacprzyk, J. (eds.) Fuzzy Systems in Medicine. STUDFUZZ, vol. 41, pp. 90–111. Physica Verlag, Heidelberg (2000)
3. Baczyński, M., Drewniak, J., Sobera, J.: Semigroups of fuzzy implications. Tatra Mt. Math. Publ. 21, 61–71 (2001)
4. Baczyński, M., Drewniak, J., Sobera, J.: On sup -∗ compositions of fuzzy implications. In: Rutkowski, L., Kacprzyk, J. (eds.) Neural Networks and Soft Computing. Advances in Soft Computing, pp. 274–279. Physica Verlag, Heidelberg (2003)
5. Baczyński, M., Jayaram, B.: Fuzzy Implications. STUDFUZZ, vol. 231. Springer, Berlin (2008)
6. Baldwin, J.F., Pilsworth, B.W.: Axiomatic approach to implication for approximate reasoning with fuzzy logic. Fuzzy Sets Syst. 3, 193–219 (1980)
7. Bandler, W., Kohout, L.J.: Fuzzy relational products as a tool for analysis and synthesis of the behaviour of complex natural and artificial systems. In: Wang, S.K., Chang, P.P. (eds.) Fuzzy Sets: Theory and Application to Policy Analysis and Information Systems, pp. 341–367. Plenum Press, New York (1980)
8. Bělohlávek, R.: Fuzzy Relational Systems. Kluwer, New York (2002)
9. Czogała, E., Drewniak, J.: Associative monotonic operations in fuzzy set theory. Fuzzy Sets Syst. 12, 249–269 (1984)
10. Drewniak, J., Kula, K.: Generalized compositions of fuzzy relations. Internat. J. Uncertain. Fuzziness Knowledge-Based Systems 10, 149–164 (2002)
11. Drewniak, J.: Invariant fuzzy implications. Soft Comput. 10, 506–513 (2006)
12. Drewniak, J., Sobera, J.: Composition of invariant fuzzy implications. Soft Comput. 10, 514–520 (2006)
13. Drewniak, J., Król, A.: A survey of weak connectives and the preservation of their properties by aggregations. Fuzzy Sets Syst. 161(2), 202–215 (2010)
14. Fodor, J.C., Roubens, M.: Fuzzy Preference Modelling and Multicriteria Decision Support. Kluwer, Dordrecht (1994)
15. Fodor, J.C., Yager, R.R., Rybalov, A.: Structure of uninorms. J. Uncertainty, Fuzzines Knowledge-Based System 5, 411–427 (1997)
16. Goguen, J.A.: L-fuzzy sets. J. Math. Anal. Appl. 18, 145–174 (1967)
17. Gottwald, S.: A Treatise on Many Valued Logics. Research Studies Press, Baldock (2001)
18. Klement, E.P., Mesiar, R., Pap, E.: Triangular Norms. Kluwer, Dordrecht (2000)
19. Łukasiewicz, J.: Interpretacja liczbowa teorii zdań. Ruch Filozoficzny 7, 92–93 (1923); english translation In: Borkowski, L. (ed.) Jan Łukasiewicz Selected Works. Studies in Logic and the Foundations of Mathematics, North Holland, Amsterdam, pp. 129–130. PWN Polish Sci. Publ., Warsaw (1970)
20. Mamdani, E.H., Assilian, S.: An experiment in linguistic synthesis with a fuzzy logic controller. Int. J. Man Machine Studies 7, 1–13 (1975)
21. Zadeh, L.A.: Fuzzy sets. Inform. Control 8, 338–353 (1965)
22. Zadeh, L.A.: Outline of a new approach to the analysis of complex systems and decision processes. IEEE Trans. Syst. Man Cyber. 3, 28–44 (1973)

Fuzzy Implications: Some Recently Solved Problems

Michał Baczyński and Balasubramaniam Jayaram

Abstract. In this chapter we discuss some open problems related to fuzzy implications, which have either been completely solved or those for which partial answers are known. In fact, this chapter also contains the answer for one of the open problems, which is hitherto unpublished. The recently solved problems are so chosen to reflect the importance of the problem or the significance of the solution. Finally, some other problems that still remain unsolved are stated for quick reference.

1 Introduction

Fuzzy implications are a generalization of the classical implication. That they form an important class of fuzzy logic connectives is clear from the fact this is the second such monograph to be exclusively devoted to them. Despite the extensive research on these operations, a few problems have remained astutely unyielding - the so called "Open Problems". A list of open problems is a particularly important source of motivation, since by exposing the inadequacies of the tools currently available, it propels the researchers towards the creation of tools or approaches that further the advancement of the topic.

In this chapter, we discuss a few of the well-known problems that have been solved, either totally or partially, since the publication of our earlier monograph [2]. The problems we deal with have not been chosen with any particular bias. However, it could be said that the choice has been dictated either based on the importance

Michał Baczyński
Institute of Mathematics,
University of Silesia, 40-007 Katowice, ul. Bankowa 14, Poland
e-mail: michal.baczynski@us.edu.pl

Balasubramaniam Jayaram
Department of Mathematics, Indian Institute of Technology Hyderabad,
Yeddumailaram - 502 205, India
e-mail: jbala@iith.ac.in

M. Baczyński et al. (Eds.): *Adv. in Fuzzy Implication Functions*, STUDFUZZ 300, pp. 177–204.
DOI: 10.1007/978-3-642-35677-3_8   © Springer-Verlag Berlin Heidelberg 2013

of the problem or the significance of the solution. This choice can also be broadly classified into two types, viz., the recently solved problems that relate to

(i) Interrelationships between the properties of fuzzy implications (Problems 1–3, Section 2),

(ii) Properties or characterizations of specific families of fuzzy implications (Problems 4–8, Section 3).

Finally, in Section 4, we also list some open problems that are yet to be solved.

2 Fuzzy Implications: Properties and Their Interrelationships

In the literature, especially in the beginning, we can find several different definitions of fuzzy implications. In this chapter we will use the following one, which is equivalent to the definition proposed by KITAINIK [25] (see also FODOR, ROUBENS [14] and BACZYŃSKI, JAYARAM [2]).

Definition 2.1. A function $I : [0,1]^2 \to [0,1]$ is called a fuzzy implication if it satisfies, for all $x, x_1, x_2, y, y_1, y_2 \in [0,1]$, the following conditions:

$$\text{if } x_1 \leq x_2, \text{ then } I(x_1, y) \geq I(x_2, y), \text{ i.e., } I(\cdot, y) \text{ is non-increasing,} \quad \text{(I1)}$$

$$\text{if } y_1 \leq y_2, \text{ then } I(x, y_1) \leq I(x, y_2), \text{ i.e., } I(x, \cdot) \text{ is non-decreasing,} \quad \text{(I2)}$$

$$I(0,0) = 1, \quad \text{(I3)}$$

$$I(1,1) = 1, \quad \text{(I4)}$$

$$I(1,0) = 0. \quad \text{(I5)}$$

While the above definition of a fuzzy implication is more or less accepted as the standard definition generalizing the classical implication operation, not all fuzzy implications possess many of the desirable properties satisfied by the classical implication on $\{0,1\}^2$ to $\{0,1\}$. Earlier definitions of fuzzy implications, assumed many of these desirable properties as part of the definition itself. For instance, TRILLAS and VALVERDE [40] also assumed the exchange principle (EP) (see Definition 2.2 below) as part of their definition of a fuzzy implication. Thus the study of the interrelationships between these properties is both interesting and imperative.

Various such properties of fuzzy implications were postulated in many works (see TRILLAS and VALVERDE [40]; DUBOIS and PRADE [11]; SMETS and MAGREZ [38]; FODOR and ROUBENS [14]; GOTTWALD [15]). The most important of them are presented below.

Definition 2.2. A fuzzy implication I is said to satisfy

(i) the exchange principle, if

$$I(x, I(y, z)) = I(y, I(x, z)), \qquad x, y, z \in [0,1]; \quad \text{(EP)}$$

(ii) the ordering property, if

$$I(x,y) = 1 \Longleftrightarrow x \le y, \qquad x,y \in [0,1]. \tag{OP}$$

The property (EP) is the generalization of the classical tautology known as the exchange principle:

$$p \to (q \to r) \equiv q \to (p \to r).$$

The ordering property (OP), called also the degree ranking property, imposes an ordering on the underlying set $[0,1]$.

2.1 Are (EP) and (OP) Sufficient?

We start with the following lemma which shows that the exchange principle (EP) together with the ordering property (OP) are strong conditions.

Lemma 2.3 (cf. [14, Lemma 1.3]). *If a function $I \colon [0,1]^2 \to [0,1]$ satisfies (EP) and (OP), then I satisfies (I1), (I3), (I4) and (I5).*

The above result shows that (EP) and (OP) force any function $I \colon [0,1]^2 \to [0,1]$ to be *almost* a fuzzy implication. The only missing property of an I satisfying (EP) and (OP) is that of (I2). However, for long, the only examples of an $I \colon [0,1]^2 \to [0,1]$ with (EP) and (OP) that satisfied (I2) was also right-continuous in the second variable. This led to the following conjecture:

Solved Problem 1 ([2, Problem 2.7.2]). *Prove or disprove by giving a counter example:*
Let $I \colon [0,1]^2 \to [0,1]$ be any function that satisfies both (EP) and (OP). Then the following statements are equivalent:

(i) I satisfies (I2).
(ii) I is right-continuous in the second variable.

One can also trace the origin of the above open problem from a different but related topic. It also arises from the characterization studies of the family of R-implications (see Definition 3.1 below and the discussion in Section 3.1).

   ŁUKASIK in [31] presented two examples (see Table 1) which finally show that the above properties are independent from each other.

2.2 Fuzzy Implication and Different Laws of Contraposition

One of the most important tautologies in the classical two-valued logic is the law of contraposition:

Table 1 The mutual independence for Problem 1

Function F		(I2)	(EP)	(OP)	Right−continuity
$F(x,y) = \begin{cases} 1, & \text{if } 0 \leq x \leq y \leq 1 \\ 1-x+y, & \text{if } 0 < y < x \leq 1 \\ 0, & \text{if } x > 0 \text{ and } y = 0 \end{cases}$		✓	✓	✓	×
$F(x,y) = \begin{cases} 1, & \text{if } 0 \leq x \leq y \leq 1 \\ \frac{1-x-3y}{1-4y}, & \text{if } 0 \leq y < x < \frac{1}{4} \\ \frac{3}{4}, & \text{if } 0 \leq y < \frac{1}{4} \leq x \leq \frac{3}{4} \\ (4y-1)x+1-3y, & \text{if } 0 \leq y < \frac{1}{4} \text{ and } \frac{3}{4} < x \leq 1 \\ y, & \text{if } \frac{1}{4} \leq y < \frac{3}{4} \text{ and } y < x \leq 1 \\ \frac{3x+y-3}{4x-3}, & \text{if } \frac{3}{4} \leq y < x \leq 1 \end{cases}$		×	✓	✓	✓

$$p \to q \equiv \neg q \to \neg p,$$

which is necessary to prove many results by contradiction. Its natural generalization to fuzzy logic is based on fuzzy negations and fuzzy implications. In fuzzy logic, contrapositive symmetry of a fuzzy implication I with respect to a fuzzy negation N (see Definition 2.4 below) plays an important role in the applications of fuzzy implications, viz., approximate reasoning, deductive systems, decision support systems, formal methods of proof, etc. (cf. [13] and [23]). Since the classical negation satisfies the law of double negation, the following laws are also tautologies in the classical logic

$$\neg p \to q \equiv \neg q \to p,$$
$$p \to \neg q \equiv q \to \neg p.$$

Consequently we can consider different laws of contraposition in fuzzy logic.

Definition 2.4 (see [14, p. 3], [27, Definition 11.3], [15, Definition 5.2.1]). A non-increasing function $N \colon [0,1] \to [0,1]$ is called a fuzzy negation if $N(0) = 1$, $N(1) = 0$. A fuzzy negation N is called

(i) strict if it is strictly decreasing and continuous;
(ii) strong if it is an involution, i.e., $N(N(x)) = x$ for all $x \in [0,1]$.

Example 2.5. The classical negation $N_C(x) = 1 - x$ is a strong negation, while $N_K(x) = 1 - x^2$ is only strict, whereas N_{D1} and N_{D2} - which are the least and largest fuzzy negations - are non-strict negations:

$$N_{D1}(x) = \begin{cases} 1, & \text{if } x = 0, \\ 0, & \text{if } x > 0, \end{cases} \qquad N_{D2}(x) = \begin{cases} 1, & \text{if } x < 1, \\ 0, & \text{if } x = 1. \end{cases}$$

Definition 2.6. Let I be a fuzzy implication and N be a fuzzy negation.

(i) We say that I satisfies the law of contraposition (or in other words, the contra-positive symmetry) with respect to N, if

$$I(x,y) = I(N(y), N(x)), \qquad x, y \in [0, 1]. \tag{CP}$$

(ii) We say that I satisfies the law of left contraposition with respect to N, if

$$I(N(x), y) = I(N(y), x), \qquad x, y \in [0, 1]. \tag{L-CP}$$

(iii) We say that I satisfies the law of right contraposition with respect to N, if

$$I(x, N(y)) = I(y, N(x)), \qquad x, y \in [0, 1]. \tag{R-CP}$$

If I satisfies the (left, right) contrapositive symmetry with respect to N, then we also denote this by CP(N) (respectively, by L-CP(N), R-CP(N)).

Firstly, we can easily observe that all the three properties are equivalent when N is a strong negation (see [2, Proposition 1.5.3]). Moreover we have the following result.

Proposition 2.7 ([2, Proposition 1.5.2]). *If $I: [0,1]^2 \to [0,1]$ is any function and N is a strict negation, then the following statements are equivalent:*

(i) I satisfies L-CP with respect to N.
(ii) I satisfies R-CP with respect to N^{-1}.

The classical law of contraposition (CP) has been studied by many authors (cf. TRILLAS and VALVERDE [40], DUBOIS and PRADE [11], FODOR [13]). It should be noted that in general it is required for N to be a strong negation and therefore it is not necessary to consider three different laws of contraposition. On the other hand, when N is only a fuzzy negation with no additional assumptions, then the different laws of contraposition may not be equivalent. In fact, only the following was known at the time of publication of [2], see Table 1.8 therein:

Table 2 Fuzzy implications and laws of contraposition

Fuzzy implication		(CP)	(L-CP)	(R-CP)
$I(x,y) = \begin{cases} \min(1, 1-x^2+y), & \text{if } y > 0, \\ 1, & \text{if } x \in [0, 0.25[\text{ and } y = 0, \\ 0.1, & \text{if } x \in [0.25, 0.75[\text{ and } y = 0, \\ 0, & \text{otherwise}. \end{cases}$		✗	✗	✗
$I_{\mathbf{YG}}(x,y) = \begin{cases} 1, & \text{if } x = 0 \text{ and } y = 0 \\ y^x, & \text{if } x > 0 \text{ or } y > 0 \end{cases}$		✗	✗	✓
$I(x,y) = \max\left(\sqrt{1-x}, y\right)$		✗	✓	✓
$I_{\mathbf{LK}}(x,y) = \min(1, 1-x+y)$		✓	✓	✓

Please note that the positive cases in Table 2 are satisfied with the natural negation of I defined by $N_I(x) := I(x,0)$, for all $x \in [0,1]$. It can be easily observed that Table 2 is not fully complete and the following question naturally arises:

Solved Problem 2 (cf. [2, Problem 1.7.1]). *Give examples of fuzzy implications I such that*

(i) I satisfies only CP(N),
(ii) I satisfies only L-CP(N),
(iii) I satisfies both CP(N) and L-CP(N) but not R-CP(N),
(iv) I satisfies both CP(N) and R-CP(N) but not L-CP(N),

with some fuzzy negation N.

BACZYŃSKI and ŁUKASIK [6] analyzed this problem and they found examples for the first two points.

Fuzzy implication		(CP)	(L-CP)	(R-CP)
$I(x,y) = \begin{cases} 0 & \text{if } x=1 \text{ and } y=0 \\ \frac{1}{7}, & \text{if } (x,y) \in \{1\}\times]0,\frac{1}{2}]\cup]\frac{1}{2},1[\times\{0\} \\ \frac{2}{7}, & \text{if } (x,y) \in]\frac{1}{2},1]\times]0,\frac{1}{2}] \\ \frac{3}{7}, & \text{if } (x,y) \in \{1\}\times]\frac{1}{2},1[\cup]0,\frac{1}{2}]\times\{0\} \\ \frac{5}{7}-\frac{1}{7}e^{-\frac{2y}{x}}, & \text{if } (x,y) \in]0,\frac{1}{2}]^2 \\ \frac{5}{7}-\frac{1}{7}e^{-\frac{2-2x}{1-y}}, & \text{if } (x,y) \in]\frac{1}{2},1[^2 \\ \frac{6}{7}, & \text{if } (x,y) \in]0,\frac{1}{2}]\times]\frac{1}{2},1[\\ 1, & \text{if } x=0 \text{ or } y=1 \end{cases}$		✓	×	×
$I'_{\mathbf{YG}}(x,y) = \begin{cases} 1, & \text{if } x=1 \text{ and } y=1 \\ (1-x)^{1-y}, & \text{otherwise} \end{cases}$		×	✓	×

Surprisingly, it is easy to show that it is not possible to find examples for next two points.

Proposition 2.8. *If a fuzzy implication I satisfies CP(N) and L-CP(N) with some fuzzy negation N, then I satisfies also R-CP(N).*

Proof. Let us fix arbitrarily $x,y \in [0,1]$. By (CP), (L-CP) and again by (CP) we get

$$I(x,N(y)) = I(N(N(y)),N(x)) = I(N(N(x)),N(y)) = I(y,N(x)),$$

so I satisfies also (R-CP) with the negation N.

In a similar way we can prove the next result.

Proposition 2.9. *If a fuzzy implication I satisfies CP(N) and R-CP(N) with some fuzzy negation N, then I satisfies also L-CP(N).*

In this way we have completely solved Problem 2.

2.3 The Law of Importation and the Exchange Principle

While the problems discussed so far arise from theoretical considerations, the problem to be discussed in this section stems from its practical significance. One of the desirable properties of a fuzzy implication, other than those listed in previous sections, is the importation law as given below:

$$I(x,I(y,z)) = I(T(x,y),z), \qquad x,y,z \in [0,1]. \tag{LI}$$

where T is a t-norm, i.e., $T: [0,1]^2 \to [0,1]$ is monotonic non-decreasing, commutative, associative with 1 as its identity element.

Fuzzy implications satisfying (LI) have been found extremely useful in fuzzy relational inference mechanisms, since one can obtain an equivalent hierarchical scheme which significantly decreases the computational complexity of the system without compromising on the approximation capability of the inference scheme. For more on this, we refer the readers to the following works [17, 39].

It can be immediately noted that if a fuzzy implication I satisfies (LI) with respect to any t-norm T, by the commutativity of the t-norm T, we have that I satisfies the exchange principle (EP).

The following problem was proposed by the authors during the Eighth FSTA conference which later appeared in the collection of such open problems by KLEMENT and MESIAR [29].

Partially Solved Problem 3 ([29, Problem 8.1]). *Let I be a fuzzy implication.*

(i) *For a given (continuous) t-norm T, characterize all fuzzy implications which satisfy the law of importation with T, i.e., the pair (I,T) satisfies (LI).*
(ii) *Since T is commutative, we know that the law of importation implies the exchange principle (EP).*

    a. *Is the converse also true, i.e., does the exchange principle imply that there exists a t-norm such that the law of importation holds?*
    b. *If yes, can the t-norm be uniquely determined?*
    c. *If not, give an example and characterize all fuzzy implications for which such implication is true.*

A first partial answer to the above question, (Problem (ii) (a)), appeared in the monograph [2, Remark 7.3.1]. As the following example shows a fuzzy implication I may satisfy (EP) without satisfying (LI) with respect to any t-norm T. Consider the fuzzy implication

$$I_{LI}(x,y) = \begin{cases} \min(1-x,y), & \text{if } \max(1-x,y) \leq 0.5, \\ \max(1-x,y), & \text{otherwise.} \end{cases}$$

If indeed there exists a T such that the above I satisfies (LI), then letting $x = 0.7, y = 1, z = 0.4$ we have

$$\text{LHS (LI)} = I_{\mathbf{LI}}(T(0.7,1),0.4) = I_{\mathbf{LI}}(0.7,0.4) = \min(1-0.7,0.4) = 0.3,$$
$$\text{RHS (LI)} = I_{\mathbf{LI}}(0.7,I_{\mathbf{LI}}(1,0.4)) = I_{\mathbf{LI}}(0.7,0) = 0 \neq 0.3.$$

Hence $I_{\mathbf{LI}}$ does not satisfy (LI) with any t-norm T.

Further, it was also shown that the t-norm T with which an I satisfies (LI) need not be unique (Problem (ii) (b)). To see this, consider the Weber implication

$$I_{\mathbf{WB}}(x,y) = \begin{cases} 1, & \text{if } x < 1 \\ y, & \text{if } x = 1 \end{cases},$$

which satisfies (LI) with any t-norm T. To see this, let $x,z \in [0,1]$. If $y = 1$, then $I_{\mathbf{WB}}(T(x,y),z) = I_{\mathbf{WB}}(x,z) = I_{\mathbf{WB}}(x,I(y,z))$. Now, let $y \in [0,1)$. Since $T(x,y) \leq y < 1$, we have $I_{\mathbf{WB}}(T(x,y),z) = 1$, and so is $I_{\mathbf{WB}}(x,I_{\mathbf{WB}}(y,z)) = I_{\mathbf{WB}}(x,1) = 1$.

Massanet and Torrens observed that though the fuzzy implication $I_{\mathbf{LI}}$ does not satisfy (LI) with any t-norm T, there exists a conjunctive commutative operator F with which it does satisfy (LI). In fact $I_{\mathbf{LI}}$ satisfies (LI) with the following uninorm

$$U(x,y) = \begin{cases} \min(x,y), & \text{if } x,y \in [0,\frac{1}{2}], \\ \max(x,y), & \text{otherwise.} \end{cases}$$

Thus they have further generalized the above problem in [34]. Note that an $F: [0,1]^2 \to [0,1]$ is said to be conjunctive if $F(1,0) = 0$.

Definition 2.10. A fuzzy implication is said to satisfy the weak law of importation if there exists a non-decreasing, conjunctive and commutative $F: [0,1]^2 \to [0,1]$ such that

$$I(x,I(y,z)) = I(F(x,y),z), \qquad x,y,z \in [0,1]. \tag{WLI}$$

It is immediately obvious that (LI) implies (WLI) which in turn implies (EP). MAS-SANET and TORRENS [34] have studied the equivalence of the above 3 properties, which has led to further interesting characterization results.

It is well-known in classical logic that the unary negation operator $\neg$ can be combined with any other binary operator to obtain the rest of the binary operators. This distinction of the unary $\neg$ is also shared by the Boolean implication $\to$, if defined in the following usual way:

$$p \to q \equiv \neg p \vee q.$$

The tautology as given above was the first to catch the attention of the researchers leading to the following class of fuzzy implications.

Definition 2.11 (see [2, Section 2.4]). A function $I: [0,1]^2 \to [0,1]$ is called an (S,N)-implication if there exist a t-conorm S and a fuzzy negation N such that

$$I(x,y) = S(N(x),y), \qquad x,y \in [0,1]. \tag{1}$$

If N is a strong fuzzy negation, then I is called a strong implication or *S-implication*. Moreover, if an (S,N)-implication is generated from S and N, then we will often denote this by $I_{S,N}$, while if N is equal to the classical negation N_C, then we will write I_S instead of I_{S,N_C}.

The following characterization of (S,N)-implications from continuous negations can be found in [2].

Theorem 2.12 ([2, Theorem 2.4.10]). *For a function $I\colon [0,1]^2 \to [0,1]$ the following statements are equivalent:*

(i) I is an (S,N)-implication with a continuous fuzzy negation N.
(ii) I satisfies (I1), (EP) and the natural negation $N_I = I(x,0)$ is a continuous fuzzy negation.

Moreover, the representation of (S,N)-implication (1) is unique in this case.

One can easily replace (EP) in the above characterization with either (WLI) or (LI). However, in this case, the mutual independence and the minimality of the properties in the above characterization need to be proven. Note that if an I satisfies (EP) and is such that N_I is continuous still it need not satisfy (I1). We know that (WLI) is stronger than (EP), a fact, that is further emphasized in [34] by the following result which proves that (WLI) and the continuity of N_I imply (I1) of I.

Proposition 2.13 ([34, Proposition 6]). *Let $I\colon [0,1]^2 \to [0,1]$ be such that it satisfies (WLI) with a non-decreasing, conjunctive and commutative function F and let N_I be continuous. Then I satisfies (I1). Hence I is a fuzzy implication, in fact, an (S,N)-implication.*

Thus we have an alternative characterization of (S,N)-implications.

Theorem 2.14 ([34, Theorem 22]). *For a function $I\colon [0,1]^2 \to [0,1]$ the following statements are equivalent:*

(i) I is an (S,N)-implication with a continuous fuzzy negation N.
(ii) I satisfies (WLI) with a non-decreasing, conjunctive and commutative function F and the natural negation N_I is a continuous fuzzy negation.

The following result plays an important role in further analysis of the above equivalences.

Lemma 2.15 (cf. [2, Lemma A.0.6]). *If N is a continuous fuzzy negation, then the function $\mathfrak{N}\colon [0,1] \to [0,1]$ defined by*

$$\mathfrak{N}(x) = \begin{cases} N^{(-1)}(x), & \text{if } x \in]0,1], \\ 1, & \text{if } x = 0, \end{cases}$$

is a strictly decreasing fuzzy negation, where $N^{(-1)}$ is the pseudo-inverse of N and is given by

$$N^{(-1)}(x) = \sup\{y \in [0,1] \mid N(y) > x\}, \qquad x \in [0,1].$$

The next two results point to the equivalence of (WLI) and (LI) when the natural negation N_I of I is continuous.

Proposition 2.16 ([34, Proposition 9]). *An (S,N)-implication obtained from a t-conorm S and a continuous fuzzy negation N satisfies* (WLI) *with the function* $F(x,y) = \mathfrak{N}(S(N(x),N(y)))$, *which is non-decreasing, conjunctive and commutative.*

Proposition 2.17 ([34, Proposition 11]). *Let I be an (S,N)-implication obtained from a t-conorm S and a continuous fuzzy negation N. Then I satisfies* (LI) *with the following t-norm T defined as*

$$T(x,y) = \begin{cases} \mathfrak{N}(S(N(x),N(y))) & \textit{if } \max(x,y) < 1, \\ \min(x,y) & \textit{if } \max(x,y) = 1. \end{cases}$$

Summarizing the above discussion, we see that two important results emerge. Firstly, we have the following result showing that both (WLI) and (LI) are equivalent in a more general setting.

Theorem 2.18 ([34, Corollary 12]). *For a function $I: [0,1]^2 \to [0,1]$ whose natural negation N_I is continuous, the following statements are equivalent:*

(i) I satisfies (WLI) *with a non-decreasing, conjunctive and commutative function F.*

(ii) I satisfies (LI) *with a t-norm T.*

Secondly, when I is a fuzzy implication whose natural negation N_I is continuous, then all of (EP), (WLI) and (LI) are equivalent.

Theorem 2.19 ([34, Proposition 13]). *Let I be a fuzzy implication whose natural negation N_I is continuous. Then the following statements are equivalent:*

(i) I satisfies (EP).

(ii) I satisfies (WLI) *with a non-decreasing, conjunctive and commutative function F.*

(iii) I satisfies (LI) *with a t-norm T.*

   Proof. (i) $\implies$ (ii): From Theorem 2.12 we know that I is an (S,N)-implication obtained from a continuous negation N and Proposition 2.16 implies that I satisfies (WLI) with a non-decreasing, conjunctive and commutative function F.

(ii) $\implies$ (iii): Follows from Theorem 2.18.

(iii) $\implies$ (i): Obvious.

Finally, the following example shows that there exist infinitely many fuzzy implications that satisfy (EP) but do not satisfy (WLI) with any non-decreasing, conjunctive and commutative function F and hence do not satisfy (LI) too.

Example 2.20 ([34, Proposition 14]). Let S be a nilpotent t-conorm, i.e., $S(x,y) = \varphi^{-1}(\min(\varphi(x) + \varphi(y), 1))$ for some increasing bijection $\varphi \colon [0,1] \to [0,1]$, and N be a strict negation. Let $I \colon [0,1]^2 \to [0,1]$ be defined as follows:

$$I(x,y) = \begin{cases} 0, & \text{if } y = 0 \text{ and } x \neq 0, \\ S(N(x),y), & \text{otherwise .} \end{cases}$$

Then I is a fuzzy implication but does not satisfy (WLI) with any non-decreasing, conjunctive and commutative function F.

3 Families of Fuzzy Implications

As already noted, fuzzy implications were introduced and studied in the literature as the generalization of the classical implication operation that obeys the following truth table:

Table 3 Truth table for the classical implication

p	q	$p \to q$
0	0	1
0	1	1
1	0	0
1	1	1

There are many ways of defining an implication in the Boolean lattice $(L, \wedge, \vee, \neg)$. Many of these have been generalized to the fuzzy context, i.e., extended as functions on $[0,1]$ instead of on $\{0,1\}$. Interestingly, the different definitions are equivalent in the Boolean lattice $(L, \wedge, \vee, \neg)$. On the other hand, in the fuzzy logic framework, where the truth values can vary in the unit interval $[0,1]$, the natural generalizations of the above definitions are not equivalent.

In the framework of intuitionistic logic the implication is obtained as the residuum of the conjunction as follows

$$p \to q \equiv \max\{t \in L \mid p \wedge t \leq q\}, \tag{2}$$

where $p, q \in L$ and the relation $\leq$ is defined in the usual way, i.e., $p \leq q$ iff $p \vee q = q$, for every $p, q \in L$. In fact, (2) is often called as the pseudocomplement of p relative to q (see [7]).

Quite understandably then, one of the most established and well-studied classes of fuzzy implications is the class of R-implications (cf. [11, 14, 15]) that generalizes the definition in (2) to the fuzzy setting.

Definition 3.1. A function $I\colon [0,1]^2 \to [0,1]$ is called an R-implication, if there exists a t-norm T such that

$$I(x,y) = \sup\{t \in [0,1] \mid T(x,t) \le y\}, \qquad x,y \in [0,1], \tag{3}$$

If an R-implication is generated from a t-norm T, then we will often denote it by I_T. Obviously, due to the monotonicity of any t-norm T, if $T(x,y) \le z$, then necessarily $x \le I_T(y,z)$. Observe that, for a given t-norm T, the pair (T, I_T) satisfies the adjointness property (also called as residual principle)

$$T(x,z) \le y \iff I_T(x,y) \ge z, \qquad x,y,z \in [0,1], \tag{RP}$$

if and only if T is left-continuous (see, for instance, the monographs [15, 2]).

Most of the early research on fuzzy implications dealt largely with these families and the properties they satisfied. In fact, still newer families of fuzzy implications are being proposed and the properties they satisfy are explored, see for instance, [35, 37].

3.1 R-implications and the Exchange Principle

From Sections 2.1 and 2.3, it is clear that (EP) and (OP) are perhaps the most important properties of a fuzzy implication both from theoretical and applicational considerations. In fact, the only characterization of R-implications are known for those that are obtained from left-continuous t-norms and both (EP) and (OP) play an important role as the result stated below demonstrates:

Theorem 3.2. *For a function $I\colon [0,1]^2 \to [0,1]$ the following statements are equivalent:*

(i) I is an R-implication generated from a left-continuous t-norm.
(ii) I satisfies (I2), (EP), (OP) and I is right continuous with respect to the second variable.

The above characterization also gave rise to many important questions. Firstly, it is necessary to answer the mutual independence and the minimality of the properties in Theorem 3.2. It is in this context that the problem discussed in Section 2.1 arose. Secondly, can a similar characterization result be obtained for R-implications generated from more general t-norms? In other words, what is the role of the left-continuity of the underlying t-norm vis-à-vis the different properties. Note that since the I considered here is not any general fuzzy implication, but whose representation is known, it is an interesting task to characterize the underlying t-norm T whose residuals satisfy the different properties stated above.

It can be shown that for any t-norm T its residual I_T satisfies (I2), while left-continuity of T is important for I_T to be right continuous with respect to the second variable. Recently, it was shown in [3] that the left-continuity of a t-norm T is not required for its residual to satisfy (OP). In fact, the following result was proven giving the equivalence between a more lenient type of continuity than left-continuity.

Definition 3.3. A function $F: [0,1]^2 \to [0,1]$ is said to be border-continuous, if it is continuous on the boundary of the unit square $[0,1]^2$, i.e., on the set $[0,1]^2 \backslash]0,1[^2$.

Proposition 3.4 ([3, Proposition 5.8], [2, Proposition 2.5.9]). *For a t-norm T the following statements are equivalent:*

(i) T is border-continuous.
(ii) I_T satisfies the ordering property (OP).

However, a similar characterization for the exchange principle, i.e., a characterization of those t-norms whose residuals satisfy (EP) is not known. Note that left-continuity of a t-norm T is sufficient for I_T to satisfy (EP), but is not necessary. Consider the non-left-continuous nilpotent minimum t-norm, which is border-continuous (see [33, p. 851]):

$$T_{\mathbf{nM}^*}(x,y) = \begin{cases} 0, & \text{if } x+y < 1, \\ \min(x,y), & \text{otherwise.} \end{cases}$$

Then the R-implication generated from $T_{\mathbf{nM}^*}$ is the following Fodor implication (Figure 1(a))

$$I_{\mathbf{FD}}(x,y) = \begin{cases} 1, & \text{if } x \le y, \\ \max(1-x,y), & \text{if } x > y, \end{cases}$$

which satisfies both (EP) and (OP). To note that (EP) and (OP) are mutually independent, consider the least t-norm, also called the drastic product, given as follows

$$T_{\mathbf{D}}(x,y) = \begin{cases} 0, & \text{if } x,y \in [0,1[, \\ \min(x,y), & \text{otherwise.} \end{cases}$$

Observe that it is a non-left-continuous t-norm. The R-implication generated from $T_{\mathbf{D}}$ is given by

$$I_{\mathbf{TD}}(x,y) = \begin{cases} 1, & \text{if } x < 1, \\ y, & \text{if } x = 1. \end{cases}$$

$I_{\mathbf{TD}}$ (see Figure 1(b)) satisfies (EP), but does not satisfy (OP). Thus the following problem appeared in [2].

Partially Solved Problem 4 ([2, Problem 2.7.3]). *Give a necessary condition on a t-norm T for the corresponding I_T to satisfy (EP).*

Note that the above problem also has relation to **Problem 4.8.1** in [2]. We will discuss this relation in detail after dealing with the solution of the above problem.

Recently, JAYARAM ET AL. [21] have partially solved the above problem for border-continuous t-norms. A complete characterization is not yet available. From the above work it can be seen that the left-continuous completion of a t-norm plays an important role in the solution. In fact, it can be seen that unless a t-norm can be embedded into a left-continuous t-norm, in some rather precise manner as presented in that work, its residual does not satisfy the exchange principle.

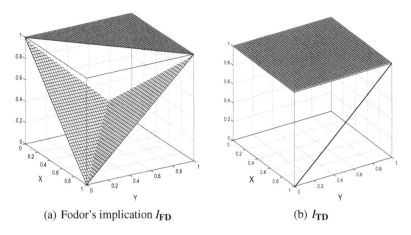

(a) Fodor's implication I_{FD} (b) I_{TD}

Fig. 1 Plots of I_{FD} and I_{TD} fuzzy implications

3.1.1 Conditionally Left-Continuous Completion

Definition 3.5. Let $F \colon [0,1]^2 \to [0,1]$ be monotonic non-decreasing and commutative. Then the function $F^* \colon [0,1]^2 \to [0,1]$ defined as below

$$F^*(x,y) = \begin{cases} \sup\{F(u,v) \mid u < x, v < y\}, & \text{if } x,y \in\,]0,1[, \\ F(x,y), & \text{otherwise,} \end{cases} \quad x,y \in [0,1], \quad (4)$$

is called the *conditionally left-continuous completion* of F.

Lemma 3.6. *If* $F \colon [0,1]^2 \to [0,1]$ *is monotonic non-decreasing and commutative, then the function* F^* *as defined in* (4) *is monotonic non-decreasing and commutative.*

Proof. By the monotonicity of F we have

$$F^*(x,y) = \begin{cases} F(x^-,y^-), & \text{if } x,y \in\,]0,1[, \\ F(x,y), & \text{otherwise,} \end{cases}$$

for any $x,y \in [0,1]$, where the value $F(x^-,y^-)$ denotes the left-hand limit. Clearly, $F^*(x,y) = F^*(y,x)$ and F^* is monotonic non-decreasing.   □

Remark 3.7. Let T be a t-norm.

(i) T^* is monotonic non-decreasing, commutative, it has 1 as its neutral element and $T^*(0,0) = 0$.
(ii) If T is border-continuous, then T^* is left-continuous (in particular it is also border-continuous).
(iii) One can easily check that I_{T^*} is a fuzzy implication.
(iv) By the monotonicity of T we have $T^* \leq T$ and hence $I_{T^*} \geq I_T$.

(v) If $x \leq y$, then $I_{T^*}(x,y) = I_T(x,y) = 1$.

(vi) Also, if $x = 1$, then by the neutrality of T we have $I_{T^*}(x,y) = I_T(x,y) = y$.

(vii) In general T^* may not be left-continuous. For example when $T = T_D$, the drastic t-norm, then $T^* = T$, but T_D is not left-continuous. This explains why T^* is called the *conditionally left-continuous completion* of T. Further, T^* may not satisfy the associativity (see Example 3.8).

Example 3.8 ([41, 42]). Consider the following non-left continuous but border-continuous Viceník t-norm given by the formula

$$T_{VC}(x,y) = \begin{cases} 0.5, & \text{if } \min(x,y) \geq 0.5 \text{ and } x+y \leq 1.5, \\ \max(x+y-1,0), & \text{otherwise.} \end{cases}$$

Then the conditionally left-continuous completion of T_{VC} is given by

$$T_{VC}^*(x,y) = \begin{cases} 0.5, & \text{if } \min(x,y) > 0.5 \text{ and } x+y < 1.5, \\ \max(x+y-1,0), & \text{otherwise.} \end{cases}$$

One can easily check that T_{VC}^* is not a t-norm since it is not associative. Indeed, we have

$$T_{VC}^*(0.55, T_{VC}^*(0.95, 0.95)) = 0.5,$$

while

$$T_{VC}^*(T_{VC}^*(0.55, 0.95), 0.95) = 0.45.$$

For the plots of both the functions see Figure 2.

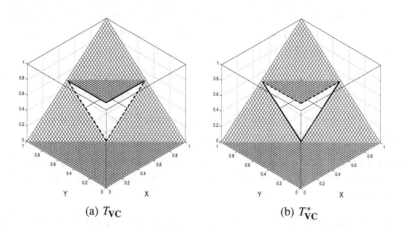

(a) T_{VC} (b) T_{VC}^*

Fig. 2 The Viceník t-norm T_{VC} and its conditionally left-continuous completion T_{VC}^* (see Example 3.8)

Definition 3.9 (cf. [24, Definition 5.7.2]). A monotonic non-decreasing, commutative and associative function $F\colon [0,1]^2 \to [0,1]$ is said to satisfy the (CLCC-A)-property, if its conditionally left-continuous completion F^*, as defined by (4), is associative.

3.1.2   Residuals of Border-Continuous T-norms and (EP)

Firstly, note that the t-norm $T_{\mathbf{B}^*}$ given below

$$T_{\mathbf{B}^*}(x,y) = \begin{cases} 0, & \text{if } x,y \in]0,0.5[, \\ \min(x,y), & \text{otherwise,} \end{cases}$$

is a border-continuous but non-left-continuous t-norm whose residual does not satisfy (EP). Indeed, the R-implication generated from $T_{\mathbf{B}^*}$ is

$$I_{\mathbf{TB}^*}(x,y) = \begin{cases} 1, & \text{if } x \le y, \\ 0.5, & \text{if } x > y \text{ and } x \in [0,0.5[, \\ y, & \text{otherwise.} \end{cases}$$

Obviously, $I_{\mathbf{TB}^*}$ satisfies (OP) but not (EP), since

$$I_{\mathbf{TB}^*}(0.4, I_{\mathbf{TB}^*}(0.5,0.3)) = 0.5,$$

while

$$I_{\mathbf{TB}^*}(0.5, I_{\mathbf{TB}^*}(0.4,0.3)) = 1.$$

The proof of main result is given in a series of lemmata. Firstly, it is shown that when T is a border-continuous t-norm and when I_T satisfies (EP), then the R-implication obtained from the conditionally left-continuous completion T^* of T is equivalen to I_T and hence also satisfies (EP).

Lemma 3.10. *Let T be a border-continuous t-norm such that I_T satisfies* (EP)*. Then $I_T = I_{T^*}$.*

Further, it is shown that, under the above assumption, T does satisfy the (CLCC-A)-property, i.e., its conditionally left-continuous completion T^* is associative.

Lemma 3.11. *Let T be a border-continuous t-norm such that I_T satisfies* (EP)*. Then T satisfies the (CLCC-A)-property, i.e., its conditionally left-continuous completion T^* is associative.*

The proof of the above result is given by showing that T^* is equal to the t-norm $T_{I_{T^*}}$ obtained from its residual I_{T^*}. Based on the above lemmata, we obtain the following partial characterization of R-implications that satisfy (EP).

Theorem 3.12. *For a border-continuous t-norm T the following statements are equivalent:*

(i) I_T satisfies (EP).
(ii) T satisfies the (CLCC-A)-property (i.e., T^* is a associative), and $I_T = I_{T^*}$.

The sufficiency follows from Lemmas 3.11 and 3.10. To see the necessity, note that if T satisfies the (CLCC-A)-property, then T^* is a left-continuous t-norm. Therefore I_{T^*} satisfies (EP). But $I_T = I_{T^*}$, so I_T also satisfies (EP).

Based on the above results a further characterization of t-norms, whose residuals satisfy both the exchange principle and the ordering property can be given as follows:

Corollary 3.13. *For a t-norm T the following statements are equivalent:*

(i) I_T satisfies (EP) and (OP).
(ii) T is border-continuous, satisfies the (CLCC-A)-property and $I_T = I_{T^*}$.

3.1.3   (EP) of an I_T and the Intersection between (S,N)- and R-implications

The exchange principle (EP) also plays an important role in determining the intersection between (S,N)- and R-implications. Many results were obtained regarding the overlaps of the above two families. Still, the following question remains and appears in the monograph [2].

Problem 4.8.1 in [2]: *Is there a fuzzy implication I, other than the Weber implication I_{WB}, which is both an (S,N)-implication and an R-implication which is obtained from a non-left continuous t-norm and cannot be obtained as the residual of any other left-continuous t-norm, i.e., is the following equality true $\mathbb{I}_{S,N} \cap (\mathbb{I}_T \setminus \mathbb{I}_{T^*}) = \{I_{WB}\}$?*

Note that for an I_T to be an (S,N)-implication, it needs to satisfy (EP) and hence the above question roughly translates into finding t-norms T such that I_T satisfies (EP).

From the results above, it is clear that when T is a border-continuous t-norm, then the above intersection is empty, i.e., $\mathbb{I}_{S,N} \cap (\mathbb{I}_T \setminus \mathbb{I}_{T^*}) = \emptyset$.

3.2   R-implications and Their Continuity

In Section 3.1 we discussed the necessity of left-continuity of a t-norm T for I_T to have certain algebraic properties, viz., (EP) and (OP), an order-theoretic property (I2) and an analytic property, that of right-continuity of I_T in the second variable.

Yet another interesting question is the continuity of I_T in both variables. Note that, since I_T is monotonic, continuity in each variable separately is also equivalent to the joint continuity of I_T in both variables. The only known continuous R-implications are those that are isomorphic to the Łukasiewicz implication, i.e., those R-implications obtained as residuals of nilpotent t-norms. In fact, these are the only known class of R-implications obtained from *left-continuous* t-norms, that are continuous. For R-implications generated from left-continuous many character-ization results are available, see for example, [14, 2]. Now we state the following main characterization result whose generalization gives the requisite answers. In the following Φ denotes the family of all increasing bijections on $[0,1]$.

Theorem 3.14 (cf. [13, Corollary 2] and [2, Theorem 2.5.33]). *For a function* $I: [0,1]^2 \to [0,1]$ *the following are equivalent:*

(i) *I is a continuous R-implication based on some left-continuous t-norm.*
(ii) *I is Φ-conjugate with the Łukasiewicz implication, i.e., there exists an increasing bijection* $\varphi: [0,1] \to [0,1]$, *which is uniquely determined, such that*

$$I(x,y) = \varphi^{-1}(\min(1 - \varphi(x) + \varphi(y), 1)), \qquad x,y \in [0,1]. \tag{5}$$

We would like to note that the proof of the above result is dependent on many other equivalence results concerning fuzzy implications, especially concerning R-implications and their contrapositivity, see, for instance, the corresponding proofs in [13, 2].

In the case of (S,N)-implications a characterization of continuous (S,N)-implica-tions was given in [1]. However, a similar complete characterization regarding the continuous subset of R-implications was not known and the following problem re-mained open for long:

Solved Problem 5 ([2, Problem 2.7.4]). *Does there exist a continuous R-impli-cation generated from non-(left)-continuous t-norm ?*

Recently, JAYARAM [18] answered the above poser in the negative, by showing that the continuity of an R-implication forces the left-continuity of the underlying t-norm and hence show that an R-implication I_T is continuous if and only if T is a nilpotent t-norm.

Before we proceed to give a sketch of this proof, let us look at some interesting consequences of the above result.

3.2.1   Importance of This Result

Firstly, using this result, one is able to resolve another question related to the inter-sections between the families of continuous R- and (S,N)-implications, which is also a generalization of an original result of SMETS and MAGREZ [38], see also [14, 15]. In particular, it can be shown that the only continuous (S,N)-implication that is also an R-implication obtained from any t-norm, not necessarily left-continuous, is the Łukasiewicz implication up to an isomorphism (see Section 3.2.3).

Note that this result also has applications in other areas of fuzzy logic and fuzzy set theory. For instance, in the many fuzzy logics based on t-norms, viz., BL-fuzzy logics [16], MTL-algebras [12] and their other variants, the negation is obtained from the t-norm itself and is not always involutive. However, the continuity of the residuum immediately implies that the corresponding negation is continuous, and hence involutive, see [5, Theorem 2.14].

It is well known that fuzzy inference mechanisms that use t-norms and their residual fuzzy implications as part of their inference scheme have many desirable properties (see, for instance, [26, 17]). Based on the results contained in this paper one can choose this pair of operations appropriately to ensure the continuity of the ensuing inference.

3.2.2 Sketch of the Proofs: Partial Functions of R-implications

We firstly note that though JAYARAM [18] answered the above problem by dealing with it exclusively, the answer could also have been obtained from some earlier works of DE BAETS and MAES [32, 8]. Interestingly, in both **the** proofs the partial functions of R-implications play an important role. In this section we detail the proof given in [18], since the proof is both independent and leads to what is perhaps - to the best of the authors' knowledge - the first independent proof of Theorem 3.14 above.

As mentioned earlier, the partial functions of R-implications play an important role. Note that since $I_T(x,x) = 1$ and $I_T(1,x) = x$, for all $x \in [0,1]$ the following definition is valid.

Definition 3.15. For any fixed $\alpha \in [0,1[$, the non-increasing partial function $I_T(\cdot,\alpha)\colon [\alpha,1] \to [\alpha,1]$ will be denoted by g_α^T.

Observe that g_α^T is non-increasing and such that $g_\alpha^T(\alpha) = 1$ and $g_\alpha^T(1) = \alpha$.

Remark 3.16. If the domain of g_α^T is extended to $[0,1]$, then this is exactly what are called *contour lines* by MAES and DE BAETS in [32, 8]. If $\alpha = 0$, then g_0^T is the natural negation associated with the t-norm T (see [2]):

$$N_T(x) = I_T(x,0) = \sup\{t \in [0,1] \mid T(x,t) = 0\}, \quad x \in [0,1].$$

In fact, the following result about these partial functions essentially states that, if the "generalized" inverse of a monotone function is continuous, then it is strictly decreasing (see [27, Remark 3.4(ii)], also [32, Theorem 11]).

Theorem 3.17. *Let T be any t-norm. For any fixed $\alpha \in [0,1[$, if g_α^T is continuous, then g_α^T is strictly decreasing.*

The rest of the proof is given in a series of Lemmata in JAYARAM [18], which we club here into a single result:

Theorem 3.18. *Let T be a t-norm such that I_T is continuous. Then*

(i) T is border continuous.
(ii) T is Archimedean, i.e., for any $x,y \in (0,1)$ there exists an $n \in \mathbb{N}$ such that $x_T^{[n]} < y$, where $x_T^{[n]} = T(x, x_T^{[n-1]})$ and $x_T^{[1]} = x$.

We note that based on Theorem 3.18 we can obtain an independent proof of Theorem 3.14. This result is based on some well-known results which we recall in the following remark.

Remark 3.19 (cf.[27]).

(i) A left-continuous T that is Archimedean is necessarily continuous and hence either strict or nilpotent (see [27, Proposition 2.16]).

(ii) A continuous Archimedean t-norm T is either strict or nilpotent.

(iii) If a continuous Archimedean t-norm T has zero divisors, then it is nilpotent (see [27, Theorem 2.18]).

(iv) A nilpotent t-norm T is a Φ-conjugate of the Łukasiewicz t-norm, i.e., there exists an increasing bijection $\varphi \colon [0,1] \to [0,1]$, which is uniquely determined, such that

$$T(x,y) = \varphi^{-1}(\max(\varphi(x) + \varphi(y) - 1, 0)), \qquad x,y \in [0,1].$$

(v) Notice that if $T(x,y) = 0$ for some $x,y \in [0,1]$, then $y \leq N_T(x)$. Moreover, if any $z < N_T(x)$, then $T(x,z) = 0$. If T is left-continuous, then $T(x,y) = 0$ for some $x,y \in [0,1]$ if and only if $y \leq N_T(x)$.

(vi) If N_T is continuous, then it is strong (see [5, Theorem 2.14]).

Corollary 3.20. *Let T be a left-continuous t-norm and I_T the R-implication obtained from it. Then the following are equivalent:*

(i) I_T is continuous.
(ii) T is isomorphic to $T_{\mathbf{LK}}$.

Proof. (i) $\Longrightarrow$ (ii): Let T be left-continuous and I_T be continuous. Then, from Theorem 3.18 above, we see that T is Archimedean and hence by Remark 3.19(i) T is necessarily continuous. Further, by Remark 3.19(ii), T is either strict or nilpotent. Now, since I_T is continuous, by Remark 3.19(v) we have that $N_T = g_0^T$ is strict and strong and hence from Remark 3.19(iv) we see that T has zero divisors. Finally, from Remark 3.19(iii), it follows that T is nilpotent and hence is isomorphic to $T_{\mathbf{LK}}$.

(ii) $\Longrightarrow$ (i): The converse is obvious, since the R-implication obtained from any nilpotent t-norm is a Φ-conjugate of the Łukasiewicz implication $I_{\mathbf{LK}}$. Since $I_{\mathbf{LK}}$ is continuous, any Φ-conjugate of it is also continuous.

Based on the above results, the main result in [18] shows that if I_T is continuous, then the left-continuity of T need not be assumed but follows as a necessity.

Theorem 3.21. *Let T be a t-norm and I_T the R-implication obtained from it. If I_T is continuous, then T is left-continuous.*

From Theorems 3.14 and 3.21 we obtain the following result.

Corollary 3.22. *For a function $I \colon [0,1]^2 \to [0,1]$ the following statements are equivalent:*

(i) I is a continuous R-implication based on some t-norm.
(ii) I is Φ-conjugate with the Łukasiewicz implication, i.e., there exists $\varphi \in \Phi$, which is uniquely determined, such that I has the form (5) for all $x,y \in [0,1]$.

3.2.3   Intersection between Continuous R- and (S,N)-implications

The intersections between the families and subfamilies of R- and (S,N)-implications have been studied by many authors, see e.g. [10, 38, 14, 2]. As regards the intersection between their continuous subsets only the following result has been known so far.

Theorem 3.23. *The only continuous (S,N)-implications that are also R-implications obtained from left-continuous t-norms are the fuzzy implications which are Φ-conjugate with the Łukasiewicz implication.*

Now, from Corollary 3.22 and Theorem 3.23 the following equivalences follow immediately:

Theorem 3.24. *For a function $I: [0,1]^2 \rightarrow [0,1]$ the following statements are equivalent:*

 (i) *I is a continuous (S,N)-implication that is also an R-implication obtained from a left-continuous t-norm.*
 (ii) *I is a continuous (S,N)-implication that is also an R-implication.*
 (iii) *I is an (S,N)-implication that is also a continuous R-implication.*
 (iv) *I is Φ-conjugate with the Łukasiewicz implication, i.e., there exists an increasing bijection $\varphi: [0,1] \rightarrow [0,1]$, which is uniquely determined, such that I has the form (5).*

3.3   Characterization of Yager's Class of Fuzzy Implications

As we have seen in earlier sections characterizations of different families of fuzzy implications are very important questions. One open problem has been connected with two families of fuzzy implications introduced by YAGER [43].

Definition 3.25. Let $f: [0,1] \rightarrow [0,\infty]$ be a strictly decreasing and continuous function with $f(1) = 0$. The function $I: [0,1]^2 \rightarrow [0,1]$ defined by

$$I(x,y) = f^{-1}(x \cdot f(y)), \qquad x,y \in [0,1], \tag{6}$$

with the understanding $0 \cdot \infty = 0$, is called an f-generated implication. The function f itself is called an f-generator of the I generated as in (6). In such a case, to emphasize the apparent relation we will write I_f instead of I.

Definition 3.26. Let $g: [0,1] \rightarrow [0,\infty]$ be a strictly increasing and continuous function with $g(0) = 0$. The function $I: [0,1]^2 \rightarrow [0,1]$ defined by

$$I(x,y) = g^{(-1)}\left(\frac{1}{x} \cdot g(y)\right), \qquad x,y \in [0,1], \tag{7}$$

with the understanding $\frac{1}{0} = \infty$ and $\infty \cdot 0 = \infty$, is called a g-generated implication, where the function $g^{(-1)}$ in (7) is the pseudo-inverse of g given by

$$g^{(-1)}(x) = \begin{cases} g^{-1}(x), & \text{if } x \in [0, g(1)], \\ 1, & \text{if } x \in [g(1), \infty]. \end{cases}$$

Based on some works of BACZYŃSKI and JAYARAM [4] on the distributive equations involving fuzzy implications, a rather not-so-elegant and partial characterization of f- and g-generated fuzzy implications can be given. However, an axiomatic characterization was unknown during the preparation of the book [2], so the following problem has been presented.

Solved Problem 6 ([2, Problem 3.3.1]). *Characterize the families of f- and g-generated implications.*

Very recently MASSANET and TORRENS [36] solved the above problem by using law of importation (LI). Firstly notice in the case when $f(0) < \infty$, the generated f-implication is an (S,N)-implication obtained from a continuous negation N (see [2, Theorem 4.5.1]) and hence the characterization result in Theorem 2.14 is applicable.

Theorem 3.27 ([36, Theorem 6]). *For a function $I : [0,1]^2 \to [0,1]$ the following statements are equivalent:*

(i) I is an f-generated implication with $f(0) < \infty$.
(ii) I satisfies (LI) with product t-norm $T_{\mathbf{P}}(x,y) = xy$ and the natural negation N_I is a continuous fuzzy negation.

When $f(0) = \infty$, then f-generated implications are not (S,N)-implications, but still similar characterizations can be proved.

Theorem 3.28 ([36, Theorem 12]). *For a function $I : [0,1]^2 \to [0,1]$ the following statements are equivalent:*

(i) I is an f-generated implication with $f(0) = \infty$.
(ii) I satisfies (LI) with product t-norm $T_{\mathbf{P}}(x,y) = xy$, I is continuous except at $(0,0)$ and $I(x,y) = 1 \Leftrightarrow x = 0$ or $y = 1$.

Similar characterizations have been obtained for g-generated fuzzy implications (see [36, Theorems 14, 17]).

3.3.1 Importance of This Result

While the (S,N)- and R-implications, dealt with in the earlier sections, are the generalizations of the material and intuitionistic-logic implications, there exists yet another popular way of obtaining fuzzy implications - as the generalization of the following implication defined in quantum logic:

$$p \to q \equiv \neg p \vee (p \wedge q).$$

Needless to state, when the truth values are restricted to $\{0,1\}$ its truth table coincides with that of the material and intuitionistic-logic implications.

Definition 3.29. A function $I: [0,1]^2 \to [0,1]$ is called a *QL-operation* if there exist a t-norm T, a t-conorm S and a fuzzy negation N such that

$$I(x,y) = S(N(x),T(x,y)), \qquad x,y \in [0,1]. \tag{8}$$

Note that not all QL-operations are fuzzy implications in the sense of Definition 2.1. A QL-operation is called a *QL-implication* only when it is a fuzzy implication. The set of all QL-implications will be denoted by $\mathbb{I}_{QL}$.

The characterization of Yager's family of f-generated implications has helped us to know the answer for two other open problems. In fact, based on the above characterization the following question, originally posed in the monograph [2] has been completely solved.

Solved Problem 7 ([2, Problem 4.8.3]).

(i) Is the intersection $\mathbb{I}_{F,\aleph} \cap \mathbb{I}_{QL}$ non-empty?
(ii) If yes, then characterize the intersection $\mathbb{I}_{F,\aleph} \cap \mathbb{I}_{QL}$.

In [36] the authors have proven the following:

Theorem 3.30 ([36, Theorem 13]). *Let $I_{T,S,N}$ be a QL-operation. Then the following statements are equivalent:*

(i) $I_{T,S,N}$ is an f-generated implication with $f(0) < \infty$.
(ii) $I_{T,S,N}$ satisfies (LI) with T_P and N is a strict negation.
(iii) N is a strict negation such that

$$N(xy) = S(N(x),T(x,N(y))), \qquad x,y \in [0,1].$$

Moreover, in this case $f = N^{-1}$, up to a multiplicative positive constant.

3.4 R-implications and a Functional Equation

The following problem was posed by HÖHLE in KLEMENT ET AL. [28]. An interesting fallout of this problem is that, as the solution shows, it gives a characterization of conditionally cancellative t-(sub)norms.

Solved Problem 8 ([28, Problem 11]). *Characterize all left-continuous t-norms T which satisfy*

$$I(x,T(x,y)) = \max(N(x),y), \quad x,y \in [0,1]. \tag{9}$$

where I is the residual operator linked to T given by (3) and

$$N(x) = N_T(x) = I(x,0), \qquad x \in [0,1].$$

Further, U. Höhle goes on to the following remark:

Remark 3.31. "*In the class of continuous t-norms, only nilpotent t-norms fulfill the above property.*"

It is clear that in the case T is left-continuous - as stated in **Problem 1** - the supremum in (3) actually becomes maximum. It is worth mentioning that the residual can be determined for more generalized conjunctions and the conditions under-which this residual becomes a fuzzy implication can be found in, for instance, [9, 22, 30]. Hence JAYARAM [19] further generalized the statement of **Problem** 8 by considering a t-subnorm instead of a t-norm and also dropping the condition of left-continuity.

Definition 3.32 ([27, Definition 1.7]). A t-subnorm is a function $M \colon [0,1]^2 \to [0,1]$ such that it is monotonic non-decreasing, associative, commutative and $M(x,y) \leq \min(x,y)$ for all $x,y \in [0,1]$.

Note that for a t-subnorm 1 need not be the neutral element, unlike in the case of a t-norm.

Definition 3.33 (cf. [27, Definition 2.9 (iii)]). A t-subnorm M satisfies the conditional cancellation law if, for any $x,y,z \in]0,1]$,

$$M(x,y) = M(x,z) > 0 \text{ implies } y = z. \tag{CCL}$$

In other words, (CCL) implies that on the positive domain of M, i.e., on the set $\{(x,y) \in (0,1]^2 \mid M(x,y) > 0\}$, M is strictly increasing. See Figure 3 (a) and (b) for examples of a conditionally cancellative t-subnorm and one that is not.

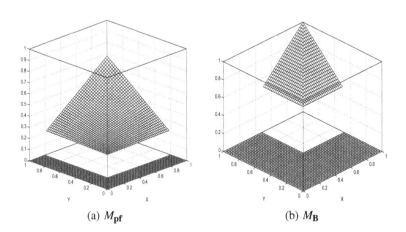

(a) $M_{\mathbf{pf}}$ (b) $M_{\mathbf{B}}$

Fig. 3 $M_{\mathbf{Pf}}$ is a conditionally cancellative t-subnorm, while $M_{\mathbf{B}}$ is not

Definition 3.34 (cf. [2, Definition 2.3.1]). Let M be any t-subnorm. Its natural negation n_M is given by

$$n_M(x) = \sup\{t \in [0,1] \mid M(x,t) = 0\}, \quad x \in [0,1].$$

Note that though $n_M(0) = 1$, it need not be a fuzzy negation, since $n_M(1)$ can be greater than 0. However, we have the following result.

Lemma 3.35 (cf. [2, Proposition 2.3.4]). *Let M be any t-subnorm and n_M its natural negation. Then we have the following:*

(i) $M(x,y) = 0 \Longrightarrow y \leq n_M(x)$.
(ii) $y < n_M(x) \Longrightarrow M(x,y) = 0$.
(iii) If M is left-continuous, then $y = n_M(x) \Longrightarrow M(x,y) = 0$, *i.e., the reverse implication of (i) also holds.*

It is interesting to note that the solution to the problem given below characterizes the set of all conditionally cancellative t-subnorms.

Theorem 3.36 ([19, Theorem 3.1]). *Let M be any t-subnorm and I the residual operation linked to M by (3). Then the following are equivalent:*

(i) The pair (I,M) satisfies (9).
(ii) M is a conditionally cancellative t-subnorm.

Example 3.37. Consider the product t-norm $T_P(x,y) = xy$, which is a strict t-norm and hence continuous and Archimedean, whose residual is the Goguen implication given by

$$I_{GG}(x,y) = \begin{cases} 1, & \text{if } x \leq y, \\ \frac{y}{x}, & \text{if } x > y. \end{cases}$$

It can be easily verified that the pair (T_P, I_{GG}) does indeed satisfy (9) whereas the natural negation of T_P is the Gödel negation

$$n_{T_P}(x) = I_{GG}(x,0) = \begin{cases} 1, & \text{if } x = 0, \\ 0, & \text{if } x > 0. \end{cases}$$

This example clearly shows that the remark of HÖHLE, Remark 3.31, is not always true. The following result gives an equivalence condition under which it is true.

Theorem 3.38 ([19, Theorem 3.2]). *Let T be a continuous t-norm that satisfies (9) along with its residual. Then the following are equivalent:*

(i) T is nilpotent.
(ii) n_T is strong.

4 Concluding Remarks

As can be seen since the publication of the monograph [2], there has been quite a rapid progress in attempts to solve open problems. However, there still remain many open problems involving fuzzy implications. In the following we list a few:

Problem 4.1. Give a necessary condition on a non-border continuous t-norm T for the corresponding I_T to satisfy (EP).

It should be mentioned that some related work on the above problem has appeared in [20]. Once again, as stated before, the above problem is also related to the following question regarding the intersection of (S,N)- and R-implications which still remains open:

Problem 4.2. (i) Is there a fuzzy implication I, other than the Weber implication $I_{\mathbf{WB}}$, which is both an (S,N)-implication and an R-implication which is obtained from a *non-border continuous* t-norm and cannot be obtained as the residual of any other left-continuous t-norm?

(ii) If the answer to the above question is in the affirmative, characterize the above non-empty intersection.

The following problems originally appeared as open in [2] and still remain so:

Problem 4.3. What is the characterization of (S,N)-implications generated from non-continuous negations?

Problem 4.4. Characterize triples (T, S, N) such that $I_{T,S,N}$ satisfies (I1).

Problem 4.5. (i) Characterize the non-empty intersection $\mathbb{I}_{\mathrm{S,N}} \cap \mathbb{I}_{\mathrm{QL}}$.

(ii) Is the Weber implication $I_{\mathbf{WB}}$ the only QL-implication that is also an R-implication obtained from a non-left continuous t-norm? If not, give other examples from the above intersection and hence, characterize the non-empty intersection $\mathbb{I}_{\mathrm{QL}} \cap \mathbb{I}_{\mathrm{T}}$.

References

1. Baczyński, M., Jayaram, B.: On the characterizations of (S,N)-implications. Fuzzy Sets and Systems 158, 1713–1727 (2007)
2. Baczyński, M., Jayaram, B.: Fuzzy Implications. STUDFUZZ, vol. 231. Springer, Berlin (2008)
3. Baczyński, M., Jayaram, B. (S,N)- and R-implications: a state-of-the-art survey. Fuzzy Sets and Systems 159, 1836–1859 (2008)
4. Baczyński, M., Jayaram, B.: On the distributivity of fuzzy implications over nilpotent or strict triangular conorms. IEEE Trans. Fuzzy Systems 17, 590–603 (2009)
5. Baczyński, M., Jayaram, B.: QL-implications: some properties and intersections. Fuzzy Sets and Systems 161, 158–188 (2010)

6. Baczyński, M., Łukasik, R.: On the different laws of contraposition for fuzzy implications. In: Proc. FSTA 2010, pp. 28–29 (2010)
7. Birkhoff, G.: Lattice Theory, 3rd edn. American Mathematical Society, Providence (1967)
8. De Baets, B., Maes, K.C.: Orthosymmetrical monotone functions. Bull. Belg. Math. Soc. Simon Stevin 14, 99–116 (2007)
9. Demirli, K., De Baets, B.: Basic properties of implicators in a residual framework. Tatra Mt. Math. Publ. 16, 1–16 (1999)
10. Dubois, D., Prade, H.: Fuzzy logic and the generalized modus ponens revisited. Internat. J. Cybernetics and Systems 15, 293–331 (1984)
11. Dubois, D., Prade, H.: Fuzzy sets in approximate reasoning. Part 1: Inference with possibility distributions. Fuzzy Sets and Systems 40, 143–202 (1991)
12. Esteva, F., Godo, L.: Monoidal t-norm based logic: towards a logic for left-continuous t-norms. Fuzzy Sets and Systems 124, 271–288 (2001)
13. Fodor, J.C.: Contrapositive symmetry of fuzzy implications. Fuzzy Sets and Systems 69, 141–156 (1995)
14. Fodor, J., Roubens, M.: Fuzzy Preference Modelling and Multicriteria Decision Support. Kluwer, Dordrecht (1994)
15. Gottwald, S.: A Treatise on Many-valued Logics. Research Studies Press, Baldock (2001)
16. Hájek, P.: Metamathematics of Fuzzy Logic. Kluwer, Dordrecht (1998)
17. Jayaram, B.: On the law of importation $(x \wedge y) \longrightarrow z \equiv (x \longrightarrow (y \longrightarrow z))$ in fuzzy logic. IEEE Trans. Fuzzy Systems 16, 130–144 (2008)
18. Jayaram, B.: On the continuity of residuals of triangular norms. Nonlinear Anal. 72, 1010–1018 (2010)
19. Jayaram, B.: Solution to an open problem: a characterization of conditionally cancellative t-subnorms. Aequat. Math. (in press), doi:10.1007/s00010-012-0143-0
20. Jayaram, B., Baczyński, M., Mesiar, R.: R-implications and the Exchange Principle: A Complete Characterization. In: Galichet, S., Montero, J., Mauris, G. (eds.) Proceedings of the 7th Conference of the European Society for Fuzzy Logic and Technology (EUSFLAT-2011) and LFA-2011, Aix-les-Bains, France, pp. 223–229 (2011)
21. Jayaram, B., Baczyński, M., Mesiar, R.: R-implications and the exchange principle: the case of border continuous t-norms. Fuzzy Sets and Systems (submitted)
22. Jayaram, B., Mesiar, R.: I-Fuzzy equivalences and I-fuzzy partitions. Inform. Sci. 179, 1278–1297 (2009)
23. Jenei, S.: A new approach for interpolation and extrapolation of compact fuzzy quantities. The one dimensional case. In: Klement, E.P., Stout, L.N. (eds.) Proc. 21th Linz Seminar on Fuzzy Set Theory, Linz, Austria, pp. 13–18 (2000)
24. Jenei, S.: A survey on left-continuous t-norms and pseudo t-norms. In: Klement, E.P., Mesiar, R. (eds.) Logical, Algebraic, Analytic and Probabilistic Aspects of Triangular Norms, pp. 113–142. Elsevier, Amsterdam (2005)
25. Kitainik, L.: Fuzzy Decision Procedures with Binary Relations. Kluwer, Dordrecht (1993)
26. Klawonn, F., Castro, J.L.: Similarity in fuzzy reasoning. Mathware and Soft Computing 2, 197–228 (1995)
27. Klement, E.P., Mesiar, R., Pap, E.: Triangular Norms. Kluwer, Dordrecht (2000)
28. Klement, E.P., Mesiar, R., Pap, E.: Problems on triangular norms and related operators. Fuzzy Sets and Systems 145, 471–479 (2004)
29. Klement, E.P., Mesiar, R.: Open problems posed at the Eight International Conference on Fuzzy Set Theory and Applications (FSTA, Liptovský Ján, Slovakia). Kybernetika 42, 225–235 (2006)

30. Król, A.: Dependencies between fuzzy conjunctions and implications. In: Galichet, S., Montero, J., Mauris, G. (eds.) Proc. EUSFLAT-LFA 2011, pp. 230–237 (2011)
31. Łukasik, R.: A note on the mutual independence of the properties in the characterization of R-implications generated from left-continuous t-norms. Fuzzy Sets and Systems 161, 3148–3154 (2010)
32. Maes, K.C., De Baets, B.: A contour view on uninorm properties. Kybernetika 42, 303–318 (2006)
33. Maes, K.C., De Baets, B.: On the structure of left-continuous t-norms that have a continuous contour line. Fuzzy Sets and Systems 158, 843–860 (2007)
34. Massanet, S., Torrens, J.: The law of importation versus the exchange principle on fuzzy implications. Fuzzy Sets and Systems 168, 47–69 (2011)
35. Massanet, S., Torrens, J.: On a new class of fuzzy implications: h-Implications and generalizations. Inform. Sci. 181, 2111–2127 (2011)
36. Massanet, S., Torrens, J.: Intersection of Yager's implications with QL and D-implications. Internat. J. Approx. Reason. 53, 467–479 (2012)
37. Massanet, S., Torrens, J.: Threshold generation method of construction of a new implication from two given ones. Fuzzy Sets and Systems 205, 50–75 (2012)
38. Smets, P., Magrez, P.: Implication in fuzzy logic. Internat. J. Approx. Reason. 1, 327–347 (1987)
39. Štěpnička, M., Jayaram, B.: On the suitability of the Bandler-Kohout subproduct as an inference mechanism. IEEE Trans. Fuzzy Systems 18, 285–298 (2010)
40. Trillas, E., Valverde, L.: On implication and indistinguishability in the setting of fuzzy logic. In: Kacprzyk, J., Yager, R.R. (eds.) Management Decision Support Systems Using Fuzzy Sets and Possibility Theory, TÜV-Rhineland, Cologne, pp. 198–212 (1985)
41. Viceník, P.: Additive generators and discontinuity. BUSEFAL 76, 25–28 (1998)
42. Viceník, P.: A note to a construction of t-norms based on pseudo-inverses of monotone functions. Fuzzy Sets and Systems 16, 15–18 (1999)
43. Yager, R.: On some new classes of implication operators and their role in approximate reasoning. Inform. Sci. 167, 193–216 (2004)

Author Index

Printed in the United States
By Bookmasters